Annals of the ICRP

ICRP PUBLICATION 114

Environmental Protection: Transfer Parameters for Reference Animals and Plants

Editor
C.H. CLEMENT

Authors on behalf of ICRP
P. Strand, N. Beresford, D. Copplestone, J. Godoy,
L. Jianguo, R. Saxén, T. Yankovich, J. Brown

PUBLISHED FOR

The International Commission on Radiological Protection

by

Los Angeles | London | New Delhi
Singapore | Washington DC

ease cite this issue as 'ICRP, 2009. Environmental Protection: Transfer Parameters for Reference Animals and Plants.
ICRP Publication 114, Ann. ICRP 39(6).'

CONTENTS

Guest Editorial

EVOLUTION OF THE ICRP SYSTEM FOR RADIATION PROTECTION OF THE ENVIRONMENT

Approaches to environmental protection, as applied internationally and nationally, take a variety of shapes and forms. This is not surprising considering the variety of aims and ambitions that govern efforts to protect the environment, which vary depending on the philosophical and ethical basis for protective actions. The variation between ecosystems and the multitude of life forms and species of those ecosystems contribute to the complexity. A number of principles have been formulated that directly and/or indirectly guide man's ambitions to protect the environment and the living organisms within it, including the pollution prevention principle, the precautionary principle, the substitution principle, and the principle of application of best-available technique.

ICRP first addressed the ethical basis for environmental protection, as well as relevant aims and principles, in *Publication 91* (ICRP, 2003). Different elements of this publication can be regarded as a precursor to the approach that ICRP has subsequently taken in terms of assessment of the effects of radiation on non-human biota, and protection of the environment from the harmful effects of radiation. It draws on experience from a number of international fora and projects, as well as national efforts, in which frameworks for assessment of radiation effects in the environment, and for protection from such effects, have been developed. *Publication 91* also explored the applicability of different approaches developed for environmental protection in general to environmental radiation protection, whilst acknowledging that any approach to protection of the environment from radiation needs to be harmonised with the system for human radiation protection; this was subject to in-depth review when *Publication 91* was in development (2003).

Publication 91 thus provided a starting point for the Commission's further considerations of how it could respond to the increased demand in society (through legislation, conventions, and accepted good practice in environmental impact assessments) to provide direct evidence of protection of the environment, as opposed to relying on the notion that actions to protect humans indirectly provide adequate protection of the environment. In this publication, ICRP stated that:

- a possible future ICRP system addressing environmental assessment and protection would focus on biota, not on the abiotic component of the environment, or on environmental media (soil, air, water, sediment);

- the system should be effect-based so that any reasoning about adequate protection would be derived from firm understanding of harm at different exposure levels; and
- the system should be based on data sets for Reference Fauna and Flora [subsequently termed 'Reference Animals and Plants' (RAPs) [1]].

The RAP approach would be analogous to the use of the Reference Person concept in human radiation protection, and would guide assessments of effects and the derivation of dose rate benchmarks to inform protective actions.

The Commission subsequently included direct consideration of environmental protection in its 2007 Recommendations (ICRP, 2007). The Commission's stated aim for environmental protection was expressed as 'preventing or reducing the frequency of deleterious radiation effects to a level where they would have negligible impact on the maintenance of biological diversity, the conservation of species, or the health and status of natural habitats, communities and ecosystems'.

In *Publication 108* (ICRP, 2008), the Commission elaborated the system for environmental protection, defined a set of 12 RAPs, and described their basic biological and life history characteristics. The RAPs were used as a basis for:

- definition of relevant exposure situations;
- development of methods to assess doses (external and internal) and derive dose conversion factors corresponding to the exposure situations; and
- analysis of effects data to generate derived consideration reference levels; these define bands of dose rates that, if observed or predicted, would trigger consideration of whether the environmental exposure under consideration gives rise to concern and possibly actions to limit the exposure.

One important data set is missing in *Publication 108*, namely concentration ratios that enable an assessor to estimate (if this is not readily measurable) the concentration of radionuclides in biota at given (measured or estimated) levels of radionuclides in the environment.

The authors of the present report have filled this gap by providing the concentration ratios for 39 radionuclides for the 12 RAPs. In doing so, they have drawn on, and contributed to, international efforts directed to generate databases for transfer factors. Reasoning and techniques to fill data gaps, where necessary, have been developed. Whilst acknowledging that there are still uncertainties as well as substantial variability, this greatly advances the practical usefulness of the RAP approach to environmental assessments and the protection of the environment from harmful effects of radiation.

[1] The definition of a RAP, as subsequently developed by the Commission in *Publication 108* (ICRP, 2008), is: 'a hypothetical entity, with the assumed basic biological characteristics of a particular type of animal or plant, as described to the generality of the taxonomic level of family, with defined anatomical, physiological, and life history properties, that can be used for the purposes of relating exposure to dose, and dose to effects, for that type of living organism'.

Whilst *Publication 108* and the current publication combined provide a robust methodology supported by a comprehensive data set, the Commission's approach to environmental protection needs to be applied sensibly and in a manner that is commensurate with the (potential) risk of harmful effects under different exposure situations. Important issues that need to be addressed in the further evolution of the Commission's system for radiation protection of the environment include:

- criteria to assist identification of situations where it would be appropriate to include environmental exposure and consideration of effects on non-human biota in assessments and/or protective actions;
- how to extrapolate from the reference data sets derived for RAPs to actual organisms of actual environments under circumstances where a detailed assessment is justified; and
- application of the system under planned, existing, and emergency exposure situations.

The issues listed here are included in the current work programme of the Commission, as well as further development, extension, and refinement of the data sets already provided. The ICRP system, or elements, are already being used to demonstrate compliance with environmental goals nationally and internationally. The guidance under development, when used in conjunction with the methodology already developed, will provide guidance to the application of the system where and when it is relevant, help identify exposure situations of concern, and provide reassurance that the environmental aspects of radiation protection have been adequately addressed in decision making.

CARL-MAGNUS LARSSON

References

ICRP, 2003. A framework for assessing the impact of ionising radiation on non-human species. ICRP Publication 91. Ann. ICRP 33 (3).

ICRP, 2007. The 2007 Recommendations of the International Commission on Radiological Protection. ICRP Publication 103. Ann. ICRP 37 (2–4).

ICRP, 2008. Environmental protection: the concept and use of reference animals and plants. ICRP Publication 108. Ann. ICRP 38 (4–6).

ICRP Publication 114

Environmental Protection: Transfer Parameters for Reference Animals and Plants

ICRP PUBLICATION 114

Approved by the Commission in April 2011

Abstract–In *Publication 103* (ICRP, 2007), the Commission included a section on the protection of the environment, and indicated that it would be further developing its approach to this difficult subject by way of a set of Reference Animals and Plants (RAPs) as the basis for relating exposure to dose, and dose to radiation effects, for different types of animals and plants. Subsequently, a set of 12 RAPs has been described in some detail (ICRP, 2008), particularly with regard to estimation of the doses received by them, at a whole-body level, in relation to internal and external radionuclide concentrations; and what is known about the effects of radiation on such types of animals and plants. A set of dose conversion factors for all of the RAPs has been derived, and the resultant dose rates can be compared with evaluations of the effects of dose rates using derived consideration reference levels (DCRLs). Each DCRL constitutes a band of dose rates for each RAP within which there is likely to be some chance of the occurrence of deleterious effects. Site-specific data on Representative Organisms (i.e. organisms of specific interest for an assessment) can then be compared with such values and used as a basis for decision making.

It is intended that the Commission's approach to protection of the environment be applied to all exposure situations. In some situations, the relevant radionuclide concentrations can be measured directly, but this is not always possible or feasible. In such cases, modelling techniques are used to estimate the radionuclide concentrations. This report is an initial step in addressing the needs of such modelling techniques.

After briefly reviewing the basic factors relating to the accumulation of radionuclides by different types of biota, in different habitats, and at different stages in the life cycle, this report focuses on the approaches used to model the transfer of radionuclides through the environment. It concludes that equilibrium concentration ratios (CRs) are most commonly used to model such transfers, and that they currently offer the most comprehensive data coverage. The report also reviews the methods used to derive CRs, and describes a means of summarising statistical information from empirical data sets. Emphasis has been placed on using data from field studies, although some data from laboratory experiments have been included for some RAPs.

There are, inevitably, many data gaps for each RAP, and other data have been used to help fill these gaps. CRs specific to each RAP were extracted from a larger database, structured in terms of generic wildlife groups. In cases where data were lacking, values from taxonomically-related organisms were used to derive suitable surrogate values. The full set of rules which have been applied for filling gaps in RAP-specific CRs is described.

Statistical summaries of the data sets are provided, and CR values for 39 elements and 12 RAP combinations are given. The data coverage, reliance on derived values, and applicability of the CR approach for each of the RAPs is discussed.

Finally, some consideration is given to approaches where RAPs and their life stages could be measured for the elements of interest under more rigorously controlled conditions to help fill the current data gaps.

Keywords: Environmental protection; Reference Animals and Plants; Concentration ratios

Authors on Behalf of ICRP

P. Strand, N. Beresford, D. Copplestone, J. Godoy, L. Jianguo, R. Saxén, T. Yankovich, J. Brown

References

ICRP, 2007. The 2007 Recommendations of the International Commission on Radiological Protection. ICRP Publication 103. Ann. ICRP 37 (2–4).

ICRP, 2008. Environmental protection: the concept and use of reference animals and plants. ICRP Publication 108. Ann. ICRP 38 (4–6).

PREFACE

Committee 5 has been systematically developing a framework, with supporting databases, in order to provide an internally consistent reference point for assessing and managing issues relating to protection of the environment. One specific and important area that was quickly identified was that of establishing a consistent means of estimating internal and external exposures in those cases where direct measurements are not possible or are unlikely to be made. A Task Group was therefore established in order to seek the best current advice on the subject, to liaise with other groups who were involved in such matters, and to provide the Commission with an up-to-date database to serve as a reference source for the ICRP's set of Reference Animals and Plants.

The membership of the Task Group was as follows:

P. Strand (Chairman)
N. Beresford
D. Copplestone
J. Godoy
L. Jianguo (from 2009)
R. Saxén (to 2009)
T. Yankovich

The corresponding member was:

J. Brown

An initial meeting was held at the Norwegian Radiation Protection Authority in April 2008, where approaches for considering transfer of radionuclides in the environment were discussed and a work programme for completing the assigned tasks was developed. A follow-up meeting was held in Monaco in June 2009 to discuss the status of the work and further issues related to the collation of transfer data through use of the Wildlife Transfer Database (http://www.wildlifetransferdatabase.org/), developed in conjunction with the International Atomic Energy Agency (IAEA).

The initial draft was discussed by the Main Commission in June 2010 and then posted on ICRP's website for consultation between July and October 2010. Many valuable comments were received and these have been taken into account in the production of the final document. The members of the Task Group are extremely grateful for the full and frank comments that were submitted. A few members of the Task Group met recently at Stirling University to address comments from the consultation and to finalise the draft report. This was resubmitted in March 2011 and cleared for publication by the Main Commission in April 2011.

The Task Group made use of the Wildlife Transfer Database when generating the transfer parameters included in this report for the ICRP Reference Animals and Plants. The Wildlife Transfer Database has been designed and supported by the Environment Agency, England and Wales; the Centre for Ecology and Hydrology

(UK); the Norwegian Radiation Protection Authority; and the Natural Environment Research Council (UK). The Task Group gratefully acknowledges the support of those organisations and individuals who have contributed to the development of the Wildlife Transfer Database. Special thanks go to the members of the IAEA Environmental Modelling for Radiation Safety II Wildlife Transfer Working Group, who are acknowledged within the IAEA Technical Report Series on wildlife transfer. The Task Group also gratefully acknowledges the constructive input of R.J. Pentreath in the drafting of this report.

The membership of Committee 5 during the production of this report was as follows:

R.J. Pentreath (Chairman) A. Real C.-M. Larsson (Vice-Chairman)
G. Pröhl K. Sakai K.A. Higley (Secretary)
P. Strand F. Brechignac D. Copplestone (from 2010)

EXECUTIVE SUMMARY

The Commission has based its approach to environmental protection on the use of a set of Reference Animals and Plants (RAPs) as the basis for relating exposure to dose, and dose to radiation effects, for different types of animals and plants in an internally consistent manner. The results of this approach have, to date, resulted in the derivation of a set of dose conversion factors for the RAPs. These dose conversion factors allow dose rates to be calculated when the concentrations of radionuclides within the RAPs and the surrounding media have been established. The resultant dose rates can then be compared with evaluations of the effects of dose rates on the different RAP types using the derived consideration reference levels outlined previously in *Publication 108* (ICRP, 2008). Each derived consideration reference level constitutes a band of dose rates for each RAP within which there is likely to be some chance of deleterious effects occurring in individuals of that type of animal or plant. Site-specific data on Representative Organisms (i.e. organisms of specific interest for an assessment) can then be compared with such values and used as a basis for decision making.

In many cases, however, direct measurements of the radionuclide concentrations in animals, plants, and the surrounding media are not available. As such, modelling techniques are often used to estimate radionuclide concentrations. Various databases have been compiled, over many years, relating to the transfer of radionuclides from environmental media to a wide range of biota, but these have been compiled primarily in order to estimate exposures to humans from their consumption. Such data therefore only usually apply to the edible parts of the relevant organisms, and to organisms that are edible. They do not, therefore, always relate to the type of organism, the life stage, or the tissue that is of interest with regard to the estimation of radiation effects. However, some data sets have been specifically derived to understand the metabolism of individual elements or radionuclides within different types of organisms; these are particularly useful but are rare.

Within this report, a number of methods that have traditionally been used to model environmental radionuclide transfer to organisms are described, and a method for deriving internal body activity concentrations in RAPs has been identified that uses empirically based concentration ratios (CRs) to relate activity concentrations in the organism to those in its environment. Equilibrium CRs are commonly used to model such transfers, and they currently offer the most comprehensive data coverage.

This report describes the formulation of a database that has allowed the collation of data on whole-body CRs and, where applicable, data entry in relation to activity concentrations in organisms and habitat media independently. For use with the RAPs, emphasis has been placed on collating data from field studies, although data from laboratory experiments have also been included for some RAPs. The database is structured in terms of generic wildlife groups, but the data have also been attributed to the RAPs where possible. In this way, CRs specifically for RAPs were extracted and, in cases where transfer data were lacking, a data-gap-filling methodology (e.g. adopting values from taxonomically-related organisms) was used to derive suitable surrogate values. The full set of rules that have been applied for filling gaps in RAP-specific CRs is described. Statistical summaries of the data sets are provided and CR values for 39 elements and 12 RAP combinations have been reported. The data coverage, reliance on derived values, and applicability of the CR approach for each of the RAPs is discussed. The results are, as to be expected, somewhat variable.

It is recognised that for radionuclides emitting relatively short-range radiations (such as alpha particles and low-energy beta radiations), and for organisms above a certain size and complexity, doses to radiosensitive tissues are likely to dictate the resultant radiation effect compared with the more commonly modelled whole-body exposure. However, few studies have been published on the internal distributions of radionuclides for many of the RAPs, and there is a lack of data on transfer for the various RAP life stages. Suggested approaches for deriving surrogate transfer data for life stages are therefore outlined.

Finally, some consideration is given to approaches where RAPs and their life stages could be measured for the elements of interest under more rigorously controlled conditions to help fill the current data gaps.

Reference

ICRP, 2008. Environmental protection: the concept and use of reference animals and plants. ICRP Publication 108. Ann. ICRP 38 (4–6).

GLOSSARY

Allometry: Relationship between the body mass of an organism and selected (physiological) parameters (e.g. radionuclide biological half-life and dietary dry matter intake).

Bioturbation: The mixing of sediment or soil by organisms, especially by burrowing or boring.

(Environmental) Compartment: A representation of a material with (relatively) homogeneous properties (e.g. soil, sediment, air, organism) created to study kinetic characteristics within a system.

Concentration ratio (CR): Activity concentration within an organism relative to that in its surrounding habitat (as represented by a particular media such as air, sediment, soil, or water).

Deposition: The process by which radionuclides are transferred from the atmosphere to the earth's surface.

Derived consideration reference level: A band of dose rate within which there is likely to be some chance of deleterious effects of ionising radiation occurring to individuals of that type of Reference Animal or Plant (derived from a knowledge of defined expected biological effects for that type of organism) that, when considered together with other relevant information, can be used as a point of reference to optimise the level of effort expended on environmental protection, dependent upon the overall management objectives and the relevant exposure situation.

Dose conversion factor: A value that enables the dose to an organism to be calculated on the assumption of a uniform distribution of a radionuclide within or external to the organism, assuming simplified dosimetry, in terms of (Gy/day)/(Bq/kg).

Distribution coefficient (K_d): The ratio of the concentrations of a radionuclide in two heterogenous phases (in this case, water and sediment) in equilibrium with each other.

ERICA: Environmental Risk from Ionising Contaminants: Assessment and Management [a European Commission/European Atomic Energy Community (EURATOM)-funded project].

Exposure: The co-occurrence or contact between the endpoint organism and the stressor(radiation or radionuclide).

Exposure assessment: The process of measuring or estimating the intensity, frequency, and duration of exposures to an agent currently present in the environment, or of estimating hypothetical exposures that might arise from the release of new chemicals into the environment.

Exposure pathway: A route by which radiation or radionuclides can reach a living organism and cause exposure.

Geochemical phase association: Speciation of radionuclides with regard to their association with various binding sites.

Redox: Oxidation–reduction reactions that describe chemical reactions in which atoms have their oxidation number (oxidation state) changed.

Reference Animal or Plant (RAP): A hypothetical entity, with the assumed basic biological characteristics of a particular type of animal or plant, as described to the generality of the taxonomic level of family, with defined anatomical, physiological, and life history properties, that can be used for the purposes of relating exposure to dose, and dose to effects, for that type of living organism.

Representative Organism: The organism or group of organisms that are the actual objects of protection in any particular assessment. In many cases, the Representative Organisms may be the same as, or very similar to, the Reference Animals and Plants, but in some cases, they may be very different.

Trophic level: A group of organisms that occupy the same position in a food chain.

1. INTRODUCTION

1.1. Aims

(1) The Commission's radiation protection framework has recently been expanded to encompass the objective of protecting the environment, the Commission having defined its aims as being those of preventing or reducing the frequency of deleterious radiation effects to a level where they would have a negligible impact on the maintenance of biological diversity; the conservation of species; or the health and status of natural habitats, communities, and ecosystems (ICRP, 2007).

(2) To achieve this objective, the Commission has decided to use a system of discrete and clearly defined Reference Animals and Plants (RAPs) for assessing radiation effects to non-human organisms, based on the concept developed by Pentreath (1998, 1999, 2002, 2004, 2005, 2009). This approach, most recently elaborated from the view of the Commission in *Publication 108* (ICRP, 2008), involves the use of a limited number of animals and plants as the basis for systematically relating exposure to dose, and then dose (or dose rate) to different types of effect, for organisms that are characteristic of different types of natural environments. A RAP is defined in ICRP (2008) as:

> *a hypothetical entity, with the assumed basic biological characteristics of a particular type of animal or plant, as described to the generality of the taxonomic level of family, with defined anatomical, physiological and life history properties, that can be used for the purposes of relating exposure to dose, and dose to effects, for that type of living organism.*

(3) The Commission reasoned that a number of RAPs were needed to reflect the variety of global operational and regulatory environmental protection requirements, as well as the need to be pragmatic in terms of developing a flexible framework to accommodate future needs and the acquisition of new knowledge. Several criteria were used to select the limited set of organism types that might be considered as typical of the terrestrial, freshwater, and marine environments (the RAPs). The use of this set of RAPs, together with their dosimetric models, underpinning data sets, knowledge about the effects of radiation, and an assessment of their relevance to wider environmental protection objectives, therefore forms the scientific basis underpinning the Commission's approach to environmental protection.

(4) Central to the approach is the intended use of these RAPs to serve as points of reference against which other data sets can be compared (Pentreath, 2005). The Commission has used this concept to develop dosimetric parameters to derive estimates of dose rate relative to external and internal concentrations of radionuclides for the different RAPs. It has also reviewed data on the effects of ionising radiation on the RAPs, and provided a set of derived consideration reference levels as a means of providing a common basis upon which decisions relating to such effects can be made (ICRP, 2008).

(5) Where radioactivity is already present in the environment, the extent to which animals and plants are exposed to radiation can be measured directly. However, for planning and other theoretical exercises, this is not the case and such exposures need to be estimated. Central to the derivation of such estimates is the need to model the transfer of radionuclides in a robust manner. An approach which could be used to estimate the internal concentrations of radionuclides in RAPs is required. The generation and utility of such data are explored in this report.

1.2. Background

(6) By way of introduction to the transfer of radionuclides in the environment, a broad overview of some of the key processes influencing the behaviour and fate of radionuclides is given below, and shown schematically in Fig. 1.1.

1.2.1. Physical and chemical processes

(7) Once released into air or water, radionuclides are influenced by physicochemical processes that lead to their dispersion in the environment. The physical and chemical form of the radionuclide, and the turbulence of the receiving medium, play an important role in relation to the initial transport mechanisms. Environmental transformation of radionuclides may also occur as radionuclides decay to daughter products, and where speciation changes over time. Other processes continually cause the transfer of contaminants from the air or the water column to the ground or sediment surface. These include the following:

- gravitational settling of suspended particulate material in atmospheric or aquatic releases (the physical size of the particulate matter is clearly an important attribute with respect to this process, as is wind speed or water velocity);
- precipitation scavenging, whereby aerosols are washed from the atmosphere by water droplets or ice crystals;
- impaction, where suspended particles impinge on solid objects within the air or water stream; and
- chemical sorption and exchange, which is dependent on the chemical and physical form of the radionuclide (i.e. speciation) and the interacting surface, and dependent on factors such as salinity, pH, and oxygen levels.

(8) Radionuclides interact with all solid materials by numerous mechanisms, including electrostatic attraction and the formation of chemical bonds. In many cases, size alone can dictate the radionuclide activity per unit mass of solid, because the surface area available for adsorption, per unit mass or volume, is greater for smaller objects. In the terrestrial environment, the interception of radionuclides by vegetation occurs by wet and dry deposition; radionuclides may also be deposited on the ground directly. Biomass per unit area can therefore affect the interception fraction for all deposition categories, but other factors, including ionic form, precipitation intensity, vegetation maturity, and leaf area index, are important when considering wet deposition (Pröhl, 2009). Radionuclide

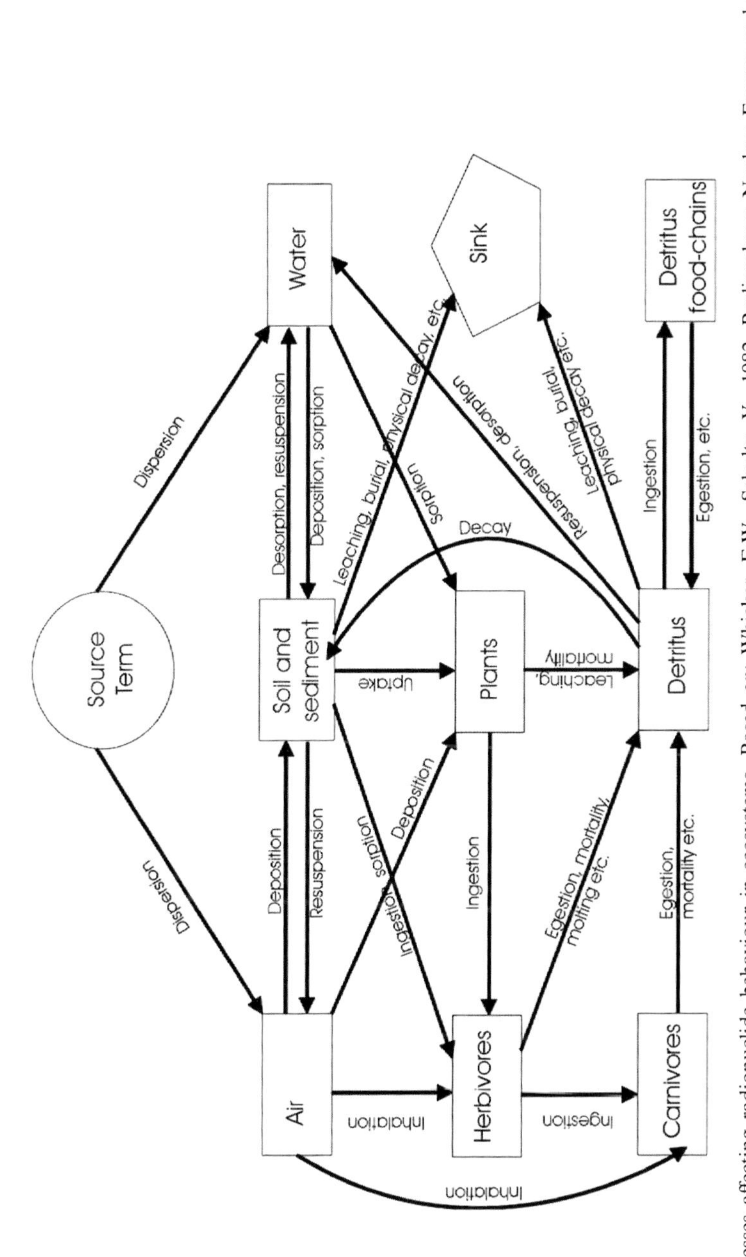

Fig. 1.1. Processes affecting radionuclide behaviour in ecosystems. Based on: Whicker, F.W., Schultz, V., 1982. Radiecology: Nuclear Energy and the Environment, Vol. 1. CRC Press, Boca Raton, FL.

concentrations on vegetation may be reduced by a number of physical processes, including wash-off by rain or irrigation, surface abrasion, leaf bending from wind action, resuspension, tissue senescence, leaf fall, herbivore grazing, and plant growth and evaporation. Various empirical formulae have been derived to model the retention of radionuclides on vegetation (IAEA, 2010).

(9) Resuspension of contaminated sediment or soil is an important process in both aquatic and terrestrial systems. In aquatic systems, turbulent action of water can resuspend surface sediments and transport them considerable distances before they are lost from the water column by sedimentation processes. Bioturbation can also be important for resuspension of particles. Such processes are important for redistributing historically contaminated sediments from open coastal sites to peripheral marine areas, where long-term sediment accumulation is occurring, such as observed by Brown et al. (1999). Furthermore, contaminated suspended sediments will be available for entry into marine food chains, especially filter-feeding organisms.

(10) In freshwater lakes, fine particulates with relatively high associated contaminant concentrations often settle in the deeper depositional areas, with coarser, less contaminated sediments found in the shallower erosional zones at the edges (Rowan et al., 1995). Such processes also occur in marine systems. In terrestrial systems, wind action and rain 'splash' on the soil can re-introduce radionuclides to the air where they can be ingested (if deposited on vegetation surfaces) or inhaled by animals. This process is influenced by factors including the height and type of the plant canopy, as well as weather (wind, rain), soil type, and animal trampling.

(11) Physical, chemical, and biological processes occurring in soil and sediment can lead to the further redistribution of radionuclides within these environmental compartments. In soils, radionuclides can migrate to deeper soil depths by, for example, leaching. Leaching rates are greatest under conditions of high rainfall and for soils containing a relatively high proportion of sand particles (Nimis, 1996). Rainfall intensity also influences leaching rates. Depending on the site-specific characteristics of the watershed, in poorly buffered surface waters, acidic snowmelt can also solubilise radionuclides, resulting in increased water concentrations at some times of the year. Upward and downward diffusional fluxes of radionuclides can result in the redistribution of contaminants within sediments, and the process of physical disturbance and bioturbation can lead to the mixing of radionuclides in the surface layer of the sediment over short time periods. Sedimentation of uncontaminated material will also lead to the long-term removal, via burial, of radionuclides. In both terrestrial and aquatic environments, animals can relocate contaminated material both horizontally and vertically through the construction of burrows, tunnels, and chambers. Plant roots can have a similar effect.

(12) The geochemical phase association of radionuclides in sediments and soils can change with time (see Vidal et al., 1993). This can affect physical transport within the ecosystem and transfer to food webs in numerous, complex ways. In some cases, a substantial proportion of the radionuclide may become associated

with residual phases, and in this way become unavailable for uptake by organisms. Such behaviour is exemplified by radiocaesium, a fraction of which can be fixed by illitic soils, the fixing process leading to virtually irreversible binding of the radionuclide to the soil matrix (Hird et al., 1996). In other cases, changes in solid-phase chemistry may lead to redistribution between geochemical phases (Bunker et al., 2000).

(13) Fractions of many radionuclides persist in exchangeable phases and, in aquatic environments, may be prone to redissolution processes whereby the contaminant is transferred from the sediment compartment to the water column, as reported by Hunt and Kershaw (1990). The fraction of a particular radionuclide present in exchangeable phases will depend on numerous factors including, amongst others, the sediment or soil characteristics, the presence of competing ions, pH, bacterial activity, and redox conditions.

1.2.2. Biological accumulation and food-chain transfer

(14) Radionuclides can enter the lowest trophic level by numerous processes. In terrestrial systems, these include direct adsorption to plant surfaces followed by foliar uptake (e.g. Zehnder et al., 1996), direct uptake via stomata (in the case of radionuclides that can be present in volatile forms, such as ^{14}C or tritium) and, more importantly for the majority of radionuclides, direct uptake via roots (or direct absorption) from soil porewater. The transfer of many radionuclides from soil to plant is thus strongly influenced by the general physical and chemical characteristics of the soil. In terrestrial systems, fungi are known to play a key role in the mobilisation, uptake, and translocation of nutrients, and are likely to contribute substantially to the long-term retention of some radionuclides in organic horizons of forest soil (Steiner et al., 2002).

(15) The transfer of radionuclides from terrestrial plants (and soil) to herbivores occurs by ingestion. Predation then leads to the transfer of radionuclides to successively higher trophic levels. When plants are consumed, they often include a soil component, which may be contaminated, adhered to the plant surface, as well as contamination incorporated within the plant itself. Radionuclides that are organically bound or present in ionic form within the plant itself may be assimilated by the herbivore to a greater degree than radionuclides that are adsorbed to soil matrices (Beresford et al., 2000). For radionuclides that are not readily taken up by plants, soil adhesion can represent the most important route of intake (IAEA, 2010). In some instances, soil ingestion by animals may be deliberate (e.g. to obtain essential minerals), but soil can also be ingested by licking or preening of fur, feathers, or offspring (Whicker and Schultz, 1982).

(16) Food webs may be very complex, with some particular food-chain pathways being very long. For example, in aquatic ecosystems, radionuclides may be either adsorbed or absorbed by bacteria, phytoplankton, and single-celled organisms, and subsequently ingested by zooplankton which can consist of an enormous variety of larval, juvenile, and adult animal forms. Due to their large surface to volume ratios, relatively high concentrations per unit weight can be achieved (e.g. Fisher et al., 1983). All of these organisms, in turn, provide food for successively higher trophic levels.

Depending on the species, aquatic primary producers can be free-floating or rooted, absorbing contaminants from the water and/or the sediments. Contaminants can then be accumulated by herbivorous and omnivorous animals that consume aquatic primary producers. The incorporation of radionuclides into sediment particles results in ingestion in various ways.

(17) Marine algae do not have roots but do have 'holdfasts' that serve to anchor them to the substrate. Radionuclides are therefore either adsorbed or absorbed directly from the water. The principal route of accumulation of radionuclides for aquatic animals is, as is the case for terrestrial animals, via ingestion. However, for some radionuclides, direct absorption from water can represent a significant proportion of the uptake. This route of uptake, in conjunction with many other transfer pathways, can be influenced by the chemistry of the ambient water, particularly in freshwater.

(18) Absorption from the gastrointestinal tract of all higher animals depends on, amongst other factors, the physicochemical form of the radionuclide, the composition of the source medium, and the nutritional status of the animal, with the radionuclides being accumulated in particular organs or body structures. Absorption is complete for some radionuclides, and can be minimal for others.

(19) The death of plants and animals, secretions, and excretions will contribute inputs of radionuclides to the detritus reservoir in terrestrial and aquatic ecosystems. Detritus can serve as an important reservoir for radionuclides through which radionuclides can be recycled back into food chains. With time, insoluble organic material, containing radionuclides, is broken down to simpler forms by the action of detritivores and, more importantly, microbes. This can lead to the release of solublised radionuclides. In contrast, deeper soil and sediment layers may act as permanent sinks for contaminants. Some of the processes discussed above, including sedimentation in the aquatic environment, leaching, and downward vertical relocation of solid material in aquatic and terrestrial systems, may lead to removal of contaminants to compartments with limited access to organisms, and biological uptake is also more limited.

(20) The kinetics of the overall system, defined by rates of transfer between environmental compartments (including soil, sediment, water, and biota groups), will determine the temporally-varying and steady-state (if attained) distribution of radionuclides within any given ecosystem. Rates of intercompartmental transport, however, vary with the radionuclides, the nature and activities of the biota, and the properties of the ecosystem.

1.2.3. Radiation exposure of biota

(21) Pathways leading to radiation exposure of plants and animals, in aquatic and terrestrial ecosystems, can be usefully considered in several different ways, as follows.

(i) Inhalation of (re)suspended contaminated particles or gaseous radionuclides (from air). This pathway is relevant for terrestrial animals and aquatic birds, reptiles, amphibians, and mammals. Respired or otherwise volatile forms of radionuclides may also contribute to the exposure of plants via gaseous exchange at the stomata.
(ii) Contamination of fur, feathers, skin, and vegetation surfaces. This has both an external exposure component (e.g. beta- and gamma-emitting radionuclides on or near the epidermis cause irradiation of the underlying living cells) and an internal exposure component (i.e. contaminants are ingested and incorporated into the body of the animal).
(iii) Ingestion of plants and animals. This leads to direct irradiation of the digestive tract, and internal exposure if the radionuclide becomes assimilated and distributed within the animal's body. For some faunal types, this will include the ingestion of detritus and sediment.
(iv) Direct uptake from the water column. This may lead to both direct irradiation of, for example, the gills or respiratory system, and internal exposure if the radionuclide becomes assimilated and distributed within the animal's body.
(v) Ingestion from water. The same exposure categories as discussed in exposure pathway (iii) are relevant here. For plants, the corresponding pathway relates to root uptake of water.
(vi) External exposure (or habitat exposure). This essentially occurs from exposure to gamma irradiation and, to a much lesser extent, beta irradiation, originating from radionuclides present in the organism's habitat. For microscopic organisms, irradiation from alpha particles may also be relevant. The configuration of the source relative to the target clearly depends on the organism's ecological characteristics and habitat. A benthic-dwelling adult fish will, for example, be exposed to radiation from radionuclides present in the water column and deposited sediments, whereas a pelagic fish may only be exposed to the former, although its eggs may be laid on or in the sediment.

(22) In the context of this report, the external irradiation arising from contamination of fur, feathers, skin, and vegetation surfaces [external component of exposure pathway (ii) in the above list] has not been considered explicitly in the derivation of transfer parameters. Accumulation of radionuclides through ingestion, and direct uptake from water pathways [exposure pathways (iii) and (iv)], have been considered in so far as they relate to internal body burdens of contaminants under (assumed) equilibrium conditions. Furthermore, the uptake of radionuclides and incorporation into the body of the organism through inhalation [exposure pathway (i)] and through the ingestion of water [exposure pathway (v)] may be indirectly included in the consideration of empirically derived transfer parameters such as CRs (as defined later) because such approaches do not differentiate between uptake routes.

(23) External exposures [exposure pathway (vi)] are not the focus of this work and are only considered in a cursory manner later in this report.

Fig. 1.2. Aquatic exposure pathways for fish and seaweed. (iii) Ingestion of animals of lower trophic levels. (iv) Direct uptake from the water column. (vi) External exposure from (a) water column and (b) sediment.

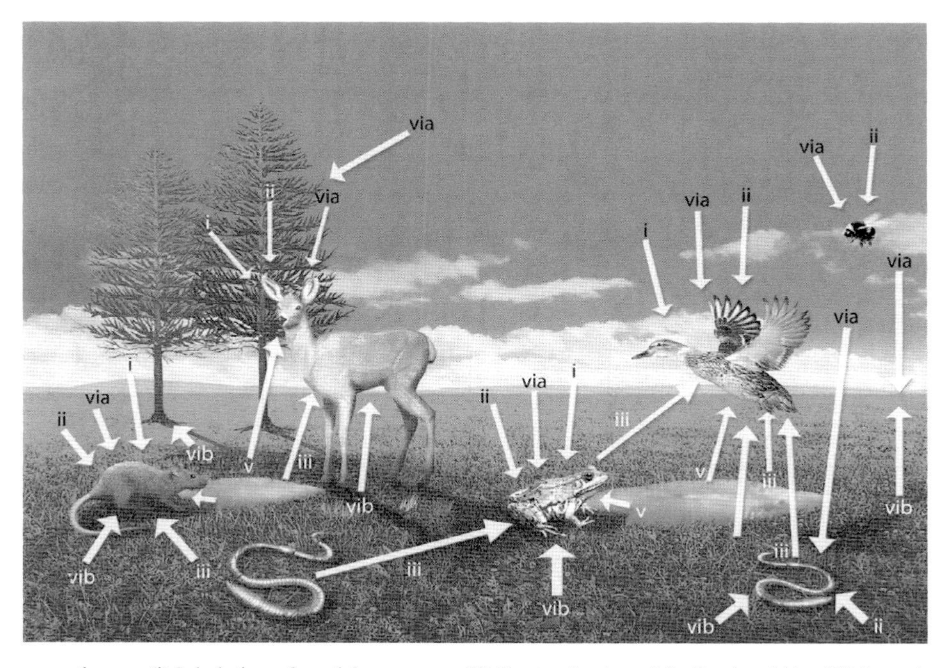

Fig. 1.3. Terrestrial exposure pathways. (i) Inhalation of particles or gases. (ii) Contamination of fur/feathers/skin. (iii) Ingestion of animals of lower trophic levels. (v) Drinking contaminated water. (vi) External exposure through (a) air or (b) soil.

(24) The exposure pathways for some aquatic and terrestrial environments are illustrated in Figs 1.2 and 1.3.

1.3. Scope

(25) This report focuses primarily on methods that allow prediction of whole-organism activity concentrations in RAPs, from a starting point of known activity concentrations of radionuclides within the organism's habitat. Modelling the physical aspects of transfer of radionuclides in the environment is beyond the scope of this work. Extensive consideration of this is reported elsewhere (e.g. IAEA, 2001, 2010). The focus of the present report is therefore on the ecological transfer of radionuclides, considering the transfer parameters that are of direct relevance, assuming that media concentrations (i.e. activity concentrations of radionuclides in water, sediment, soil, or air) are available from either direct measurement or from appropriate model simulations.

(26) The radionuclides considered (Table 1.1) were selected by Committee 5 of the ICRP and used to provide dose conversion factors for the RAPs (ICRP, 2008).

(27) The Commission intends that its approach to environmental protection should apply to all exposure situations that it considers. These are as follows.

- Planned exposure situations: everyday situations involving the planned operation of sources including decommissioning, disposal of radioactive waste, and rehabilitation of the previously occupied land. Practices in operation are planned exposure situations.
- Existing exposure situations: situations that already exist when a decision on control has to be taken, including natural background radiation and residues from past practices that were operated outside the Commission's recommendations.
- Emergency exposure situations: unexpected situations that occur during the operation of a practice, requiring urgent action. Emergency exposure situations may arise from practices.

(28) For the sake of simplicity, and given the intention to be as broadly applicable as possible, a decision was made to focus on approaches that are appropriate under equilibrium or quasi-equilibrium conditions. These are essentially the conditions that might be expected to exist where the environment is receiving continuous inputs of radionuclides from facilities operating under a regulated discharge regime, or at historically contaminated sites where inputs have ceased. The transfer parameter values provided in this report are applicable to planned and existing exposure situations that are in equilibrium, and they are less suitable for evolving emergency exposure situations. This, however, depends on the time scales involved. Thus, for exposures at some stages of the life cycle, such as for eggs and larvae, where the life stage is only a matter of days, there may be little difference.

(29) In *Publication 108* (ICRP, 2008), the Commission described how RAPs may be used to derive numerical values to enable managerial action to be taken within the radiological protection framework when environmental assessments are required. This is outlined in Fig. 1.4. The approach parallels the ICRP's system of radiological

Table 1.1. Elements and their radioisotopes considered in this report.

Element		Isotopes
Ag	Silver	Ag-110m
Am	Americium	Am-241
Ba	Barium	Ba-140
C	Carbon	C-14
Ca	Calcium	Ca-45
Cd	Cadmium	Cd-109
Ce	Cerium	Ce-141, Ce-144
Cf	Californium	Cf-252
Cl	Chlorine	Cl-36
Cm	Curium	Cm-242, Cm-243, Cm-244
Co	Cobalt	Co-57, Co-58, Co-60
Cr	Chromium	Cr-51
Cs	Caesium	Cs-134, Cs-135, Cs-136, Cs-137
Eu	Europium	Eu-152, Eu-154
H	Tritium	H-3
I	Iodine	I-125, I-129, I-131, I-132, I-133
Ir	Iridium	Ir-192
K	Potassium	K-40
La	Lanthanum	La-140
Mn	Manganese	Mn-54
Nb	Niobium	Nb-94, Nb-95
Ni	Nickel	Ni-59, Ni-65
Np	Neptunium	Np-237
P	Phosphorus	P-32, P-33
Pa	Protactinium	Pa-231
Pb	Lead	Pb-210
Po	Polonium	Po-210
Pu	Plutonium	Pu-238, Pu-239, Pu-240, Pu-241
Ra	Radium	Ra-226, Ra-228
Ru	Ruthenium	Ru-103, Ru-106
S	Sulphur	S-35
Sb	Antimony	Sb-124, Sb-125
Se	Selenium	Se-75, Se-79
Sr	Strontium	Sr-89, Sr-90
Tc	Technetium	Tc-99
Te	Tellurium	Te-129m, Te-132
Th	Thorium	Th-227, Th-228, Th-230, Th-231, Th-232, Th-234
U	Uranium	U-234, U-235, U-238
Zn	Zinc	Zn-65
Zr	Zirconium	Zr-95

protection for humans in terms of intake and external exposure. Within this system, various data sets relating specifically to parameters including anatomy, physiology, and dosimetry allow effective doses to a 'Reference Person' to be derived, and dose limits, dose constraints, and reference levels to be established. This is the role that the RAPs will undertake, and this report represents one of the building blocks to enable the RAPs to be used in this way.

(30) However, when it comes to demonstrating compliance, the Commission recommends that an assessment should consider the exposure to a Representative

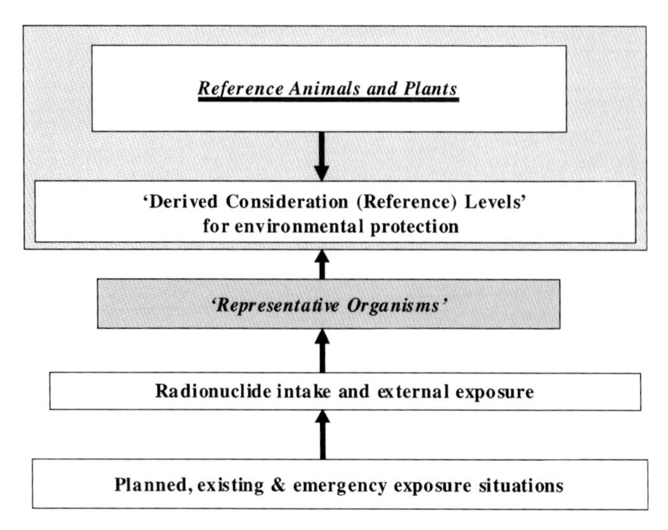

Fig. 1.4. Relationships between various points of reference for protection of the environment. Source: ICRP (2008). Environmental protection: the concept and use of reference animals and plants. ICRP Publication 108. Ann. ICRP 38(4–6).

Person for human radiological protection. The Representative Person may be real or hypothetical, but the habits used (e.g. consumption of foodstuffs, location, use of local resources) should be typical of those of a small number of individuals who are most highly exposed. As shown in Fig. 1.4, the Commission recognises that a set of Representative Organisms could similarly be defined for environmental assessments of non-human species, and their doses could be evaluated to reflect the exposure situation being considered more realistically.

(31) The values compiled in this report are intended to be a data set that helps to explore the relationships specifically between activity concentrations in RAPs and their habitats. These data should not, therefore, be considered as replacement values to be used instead of site-specific or species-specific data; for example, where possible, measured activity concentrations or transfer parameters for the particular plants and animals, i.e. Representative Organisms, within a specific site under assessment should still be used. They are, instead, intended to provide the Commission with values to use when exploring the relationship between dose and biological effect. These values may be compared with site-specific transfer parameter values obtained for the Representative Organisms, or could be used as surrogates when other data are lacking for particular assessments. The databases on exposure, dosimetry, background irradiation, and biological effects that the Commission is constructing around the RAPs will provide the basic tools for performing an internally consistent exposure analysis. An assessor will therefore be able to provide a risk characterisation explicitly for RAPs at a particular defined site. By normalising the assessment in this way, the quality and robustness of the analysis might be checked and contextualised through comparison with assessments for RAPs conducted elsewhere. Future publications will address how RAPs and Representative Organisms may be applied to different exposure situations.

1.4. References

Beresford, N.A., Mayes, R.W., Cooke, A.I., et al., 2000. The importance of source dependent bioavailability in determining the transfer of ingested radionuclides to ruminant derived food products. Environ. Sci. Technol. 34, 4455–4462.

Brown, J.E., McDonald, P., Parker, A., et al., 1999. Specific activity profiles with depth in a Ribble Estuary saltmarsh: interpretation in terms of radionuclide transport and dispersion mechanisms in the marine and estuarine environments of NW England. J. Environ. Radioact. 43, 259–275.

Bunker, D.J., Smith, J.T., Livens, F.R., et al., 2000. Determination of radionuclide exchangeability in freshwater systems. Sci. Total Environ. 263, 171–183.

Fisher, N.S., Bjerregaard, P., Fowler, S.W., 1983. Interactions of marine plankton with transuranic elements. 1. Biokinetics of neptunium, plutonium, americium and californium in phytoplankton. Limnol. Oceanogr. 28, 432–447.

Hird, A.B., Rimmer, D.L., Livens, F.R., et al., 1996. Factors affecting the sorption and fixation of caesium in acid organic soils. Eur. J. Soil Sci. 47, 97–104.

Hunt, G.J., Kershaw, P.J., 1990. Remobilisation of artificial radionuclides from the sediment of the Irish Sea. J. Radiol. Prot. 10, 147–151.

IAEA, 2001. Generic Models for Use in Assessing the Impact of Discharges of Radioactive Substances to the Environment. Safety Report Series No. 19. International Atomic Energy Agency, Vienna.

IAEA, 2010. Handbook of Parameter Values for the Prediction of Radionuclide Transfer in Terrestrial and Freshwater Environments. IAEA Technical Report Series No. 472. International Atomic Energy Agency, Vienna.

ICRP, 2007. The 2007 Recommendations of the International Commission on Radiological Protection. ICRP Publication 103. Ann. ICRP 37(2–4).

ICRP, 2008. Environmental protection: the concept and use of reference animals and plants. ICRP Publication 108. Ann. ICRP 38(4–6).

Nimis, P.L., 1996. Radiocesium in plants of forest ecosystems. Stud. Geobotan. 15, 3–49.

Pentreath, R.J., 1998. Radiological protection criteria for the natural environment. Radiat. Prot. Dosim. 75, 175–179.

Pentreath, R.J., 1999. A system for radiological protection of the environment: some initial thoughts and ideas. J. Radiol. Prot. 19, 117–128.

Pentreath, R.J., 2002. Radiation protection of people and the environment: developing a common approach. J. Radiol. Prot. 22, 1–12.

Pentreath, R.J., 2004. Ethics, genetics and dynamics: an emerging systematic approach to radiation protection of the environment. J. Environ. Radioact. 74, 19–30.

Pentreath, R.J., 2005. Concept and use of reference animals and plants. In: Protection of the Environment from the Effects of Ionising Radiation, 6-10 October 2003, Stockholm. IAEA-CN-109. International Atomic Energy Agency, Vienna, pp. 411–420.

Pentreath, R.J., 2009. Radioecology, radiobiology, and radiological protection: frameworks and fractures. J. Environ. Radioact. 100, 1019–1026.

Pröhl, G., 2009. Interception of dry and wet deposited radionuclides by vegetation. J. Environ. Radioact. 100, 675–682.

Rowan, D.J., Cornett, R.J., King, K., et al., 1995. Sediment focusing and Pb-210 dating: a new approach. J. Paleolimnol. 13, 107–118.

Steiner, M., Linkov, I., Yoshida, S., 2002. The role of fungi in the transfer and cycling of radionuclides in forest ecosystems. J. Environ. Radioact. 58, 217–241.

Vidal, M., Tent, J., Llaurado, M., et al., 1993. Study of the evolution of radionuclide distribution in soils using sequential extraction schemes. J. Radioecol. 1, 49–55.

Whicker, F.W., Schultz, V., 1982. Radiecology: Nuclear Energy and the Environment, Vol. 1. CRC Press, Boca Raton, FL.

Zehnder, H-J., Kopp, P., Eikenberg, J., et al., 1996. Uptake and transport of radioactive cesium and strontium into strawberry plants and grapevines after leaf contamination. In: Desmet, G., Howard, B.J., Heinrich, G., Schimmack, W. (Eds.), Proceedings of the International Symposium on Radioecology, 22–24 April 1996, Vienna, pp. 155–161. Austrian Soil Science Society & Federal Environment Agency, Vienna.

2. OVERVIEW OF APPROACHES USED TO MODEL TRANSFER OF RADIONUCLIDES IN THE ENVIRONMENT

(32) A number of approaches have been proposed, in the context of conducting exposure assessments, to estimate transfer of radionuclides to biota when measured activity concentrations are not available. These range from tabulated transfer parameters (e.g. Brown et al., 2003) to integrated approaches that employ spreadsheets incorporating transfer data (e.g. Copplestone et al., 2001, 2003; Brown et al., 2008), and more highly parameterised food-chain models (Thomann, 1981; Brown et al., 2004; USDOE, 2004).

2.1. Concentration ratios

(33) There are several definitions of concentration ratio (CR). The CR for the different ecosystems considered is defined here as follows.

(34) For terrestrial biota,

$$CR = \frac{A_{b,r}^{biota}}{A_{r}^{soil}} \tag{2.1}$$

where $A_{b,r}^{biota}$ is the activity concentration of radionuclide r in whole-body biota b (Bq/kg fresh weight), and A_{r}^{soil} is the activity concentration of radionuclide r in soil (Bq/kg dry weight).

(35) For some approaches, exceptions exist for chronic atmospheric releases of ^{3}H, $^{32,33}P$, ^{35}S, and ^{14}C where:

$$CR = \frac{A_{b,r}^{biota}}{A_{r}^{air}} \tag{2.2}$$

where $A_{b,r}^{biota}$ is the activity concentration of radionuclide r in whole-body biota b (Bq/kg fresh weight), and A_{r}^{air} is the activity concentration of radionuclide r in air (Bq/m^3).

(36) For aquatic biota,

$$CR - \frac{A_{b,r}^{biota}}{A_{r}^{water}} \tag{2.3}$$

where $A_{b,r}^{biota}$ is the activity concentration of radionuclide r in whole-body biota b (Bq/kg fresh weight), and A_{r}^{water} is the activity concentration of radionuclide r in (normally filtered) water (Bq/l).

(37) The CR approach is simple, based on empirical data, and combines (empirically) all the various transfer pathways (e.g. radionuclide intake via food, soil ingestion, inhalation, and drinking water in terrestrial animals). Furthermore, all existing approaches to estimate the exposure of wildlife use CR values, as defined above, for at least some organisms. Determination and application of CR values are, however, subject to factors such as sampling methodology, the degree of equilibrium between

biota and media, and environmental parameters such as water chemistry and soil type (see Beresford et al., 2004; Yankovich et al., 2010), although the alternative approaches discussed below are also subject to many of these factors. With respect to water chemistry, some models propose simple relationships between water–stable element concentrations in water and radionuclide transfer to biota (e.g. Smith, 2006; Yankovich et al., 2010).

(38) The most comprehensive recent review of CRs, based on the concept of generic wildlife groups (termed 'Reference Organisms'), was conducted as part of the ERICA project (Larsson, 2008). In this respect, Beresford et al. (2008a) and Hosseini et al. (2008) presented a complete set of CR values for more than 1100 radionuclide–organism combinations in terrestrial, freshwater, and marine ecosystems. By preference, CR values were derived from reviews of original publications including the use of stable element data.

2.2. Alternative approaches used in quantifying radionuclide transfer

(39) Some models use alternative approaches to determine the transfer of radionuclides to birds and mammals. For example, to provide transfer parameters when CR values are lacking, USDOE (2002) suggests a kinetic–allometric approach to predicting radionuclide concentrations in animals. Allometry, or 'biological scaling', is the consideration of the effect of mass on biological variables. The dependence of a biological variable, Y, on a body mass, M, is typically characterised by allometric equations of the form:

$$\mathrm{Y} = a\mathrm{M}^{b} \tag{2.4}$$

where a and b are constants.

(40) There are a number of publications summarising allometric relationships for a wide range of biological variables (e.g. Hoppeler and Weibel, 2005). Many biological phenomena appear to scale as quarter powers of the mass (Brown et al., 2000; West et al., 2000). For example, metabolic rates scale as $\mathrm{M}^{0.75}$, rates of cellular metabolism and maximal population growth rate as $\mathrm{M}^{-0.25}$, lifespan and embryonic growth and development as $\mathrm{M}^{0.25}$, and cross-sectional areas of mammalian aortas and tree trunks as $\mathrm{M}^{0.75}$. Allometric relationships for the biological half-life and dietary transfer coefficient for some radionuclides have been derived by a number of authors, and most of these coefficients also scale to quartile values (see Beresford et al., 2004).

(41) MacDonald (1996) derived allometric relationships describing the transfer of caesium and iodine from feed to the tissues of wild mammalian and bird species which scaled to approximately −0.7. Since then, USDOE (2002) has provided biological half-lives for 15 elements which can be used, together with allometric relationships, to derive daily dietary intake, water intake, inhalation rates, and parameters describing soil/sediment ingestion and gastrointestinal absorption to estimate whole-body activity concentrations for terrestrial and riparian mammals and birds. In recent intercomparison exercises, allometric models have been demonstrated to give results comparable with CR value parameterised approaches (Beresford et al.,

2008c). Application of allometric models to marine mammals was proposed by Brown et al. (2005), and to marine species generally by Vives i Batlle et al. (2007).

(42) Using algebraic derivations, and the allometric relationships for radionuclide biological half-lives or transfer coefficients, and dietary dry matter intake, Beresford et al. (2004) proposed that, for many radionuclides, the biota-to-dietary CR would be constant across species. This has been used subsequently to provide some transfer parameters for assessment models (Beresford et al., 2008a).

(43) Some models have attempted to provide a complete set of transfer parameters for the radionuclides and organisms they consider. As discussed above within US-DOE (2002), this was achieved by the development of allometric approaches. Data were only available for approximately 40% of the >1100 CR values required for the default transfer database of the ERICA tool (Brown et al., 2008). Consequently, a set of options was established (Beresford et al., 2008a) which represented an evolution of the approach initially proposed by Copplestone et al. (2003). The options were as follows.

- Use an available CR value for an organism of similar taxonomy within a given ecosystem for the radionuclide under assessment (preferred option).
- Use an available CR value for a similar Reference Organism within a given ecosystem for the radionuclide under assessment (preferred option).
- Use an available CR value for the given Reference Organism for an element of similar biogeochemistry.
- Use an available CR value for biogeochemically similar elements for organisms of similar taxonomy.
- Use an available CR value for biogeochemically similar elements available for a similar Reference Organism.
- Use allometric relationships, or other modelling approaches, to derive appropriate CRs.
- Assume the highest available CR (least preferred option).
- Use the CR for the same organism in a different ecosystem (least preferred option).

(44) Further details concerning the application of these options are provided in Beresford et al. (2008a) and Hosseini et al. (2008) for terrestrial and aquatic ecosystems, respectively.

(45) A number of dynamic models have been proposed for use in assessing exposure of terrestrial (e.g. Avila et al., 2004), freshwater (see Beresford et al., 2008c), and marine (e.g. Vives i Batlle et al., 2007, 2008) biota. Some of these are adaptations of models originally proposed to predict radionuclide contamination of human foodstuffs. For dynamic or biokinetic models, transfer from the environment to plants and animals is modelled as a time-dependent function that can take into account variations in environmental activity concentrations over time. Typically, such models are characterised by discrete compartments representing particular abiotic and biotic components within the environment, and with transfer from or between compartments being described by rate constants, e.g. rates characterising biological half-lives of uptake and elimination.

(46) For some radionuclide–organism combinations, comparison of the available models, presented above as CRs and alternative approaches, has demonstrated significant (orders of magnitude) variation in biota activity concentration predictions (Beresford et al., 2008c).

(47) In aquatic ecosystems, most approaches make use of distribution coefficients (K_d) to describe the relative activity concentrations in sediment compared with water. The K_d value is used to estimate sediment concentrations from water concentrations or vice versa if data for either are lacking. Whilst biota activity concentrations are determined in aquatic ecosystems from those in water, sediment concentrations are required to estimate external dose rates. Although the application of distribution coefficients forms an integral part of many exposure assessments, the concept and application of such models is not unique to RAPs. The collation and derivation of statistical information and representative values for sediment distribution coefficients has been the subject of comprehensive reviews elsewhere (IAEA, 2004, 2010), and the reader is referred to these compilations for further details.

2.3. Selection of approach to provide baseline transfer parameters for the ICRP Reference Animals and Plants

(48) In *Publication 108* (ICRP, 2008), the Commission considered radionuclides for 40 elements and 12 RAPs with their associated life stages. A number of data sets are available which can be used to provide transfer parameter values for the RAPs.

(49) The CR value databases developed for the ERICA tool (Brown et al., 2008) considered Reference Organisms which encompass all of the adult stages of the RAPs, although only limited data for their other life stages. The ERICA tool contains information for 31 of the 40 elements given in Table 1.1. This represents a broader coverage for the RAPs than any of the other approaches/databases discussed previously. With some exceptions, the ERICA tool has also given reasonable predictions of the internal activity concentrations when applied at sites for which biota activity concentration measurements were available, and generally compares favourably against other approaches (Beresford et al., 2007, 2008b,c, 2010; Wood et al., 2008; Yankovich et al., 2010). The ERICA transfer databases incorporated all sources used in some previous reviews (including Copplestone et al., 2001; Brown et al., 2003) and some source references used by USDOE (2002).

(50) From a pragmatic perspective, CR values are simple to apply, represent the most comprehensive database available, and the methodology is analogous to approaches used for some aspects of human radiological assessments (e.g. IAEA, 2010). Currently, the Commission considers the CR approach to provide a reasonable starting point from which to explore the transfer of radionuclides to the RAPs.

(51) However, there are some caveats to application of the CR approach. CR values represent an amalgamation and simplification of many transfer processes, and most suitably represent long-term, average, steady-state conditions. The application of (equilibrium) CR values is not appropriate to highly dynamic scenarios such as those that may follow an accidental release. Nevertheless, rapidly changing dynamic situations such as those seen in an emergency will, eventually, without a sharp

borderline in time, transform into an existing exposure situation, where the use of equilibrium transfer models may be more robustly justified.

(52) Many radionuclides will be deposited and retained internally within organisms, sometimes over very long time scales. It has been assumed for humans, by way of example, that plutonium deposited in liver has a biological half-life of 20 years, and plutonium deposited in bone has a biological half-life of 50 years (ICRP, 1988). Using such protracted retention times in biokinetic models essentially results in no equilibrium being attained during the lifetime of the (human) individual, and for a constant ingestion rate of this actinide, the body burden simply increases with time. A similar situation is to be expected for some of the longer-lived organisms considered here.

(53) The CR approach therefore provides a pragmatic and proportionate approach to identifying the internal body activity concentration for use in the assessment of radiological impact on non-human species. While it is theoretically possible to adopt a similar approach for non-human species as that used for human assessments, including the compilation of data on physiology, form, and structure of the body, plus elemental composition of the tissues and organs, as was done for Reference Man (ICRP, 1975), this level of detail may be too great for some organisms relative to the known biological effects data, and the derived consideration reference levels and dose conversion factor values presented by the Commission (ICRP, 2008).

2.4. References

Avila, R., Beresfold, N., Broed, R., et al., 2004. Study of the uncertainty in estimation of the exposure of non-human biota to ionizing radiation. J. Radiol. Prot. 24, A105–A122.

Beresford, N.A., Broadley, M.R., Howard, B.J., et al., 2004. Estimating radionuclide transfer to wild species – data requirements and availability for terrestrial ecosystems. J. Radiol. Prot. 24, A89–A103.

Beresford, N.A., Howard, B.J., Barnett, C.L., 2007. Application of ERICA Integrated Approach at Case Study Sites. Deliverable 10 of the ERICA Project (FI6R-CT-2004-508847). CEH-Lancaster, Lancaster.

Beresford, N.A., Barnett, C.L., Howard, B.J., et al., 2008a. Derivation of transfer parameters for use within the ERICA tool and the default concentration ratios for terrestrial biota. J. Environ. Radioact. 99, 1393–1407.

Beresford, N.A., Gaschak, S., Barnett, C.L., et al., 2008b. Estimating the exposure of small mammals at three sites within the Chernobyl exclusion zone – a test application of the ERICA tool. J. Environ. Radioact. 99, 1496–1502.

Beresford, N.A., Barnett, C.L., Brown, J., et al., 2008c. Inter-comparison of models to estimate radionuclide activity concentrations in non-human biota. Radiat. Environ. Biophys. 47, 491–514.

Beresford, N.A., Barnett, C.L., Brown, J.E., et al., 2010. Predicting the radiation exposure of terrestrial wildlife in the Chernobyl exclusion zone: an international comparison of approaches. J. Radiol. Prot. 30, 341–373.

Brown, J.H., West, G.B., Enquist, B.J., 2000. Patterns and processes, causes and consequences. In: Brown, J.H., West, G.B. (Eds.), Scaling in Biology. Oxford University Press, Oxford, pp. 1–24.

Brown, J.E., Strand, P., Hosseini, A., et al., 2003. Handbook for Assessment of the Exposure of Biota to Ionising Radiation from Radionuclides in the Environment. Deliverable Report for the EC Project FASSET (Contract No. FIGE-CT-2000-00102). Norwegian Radiation Protection Authority, Østerås.

Brown, J., Børretzen, P., Dowdall, M., et al., 2004. The derivation of transfer parameters in the assessment of radiological impacts to Arctic marine biota. Arctic 57, 279–289.

Brown, J.E., Børretzen, P., Hosseini, A., 2005. Biological transfer of radionuclides in marine environments – identifying and filling knowledge gaps for environmental impact assessments. Radioprotection 40, 533–539.

Brown, J.E., Alfonso, B., Avila, R., et al., 2008. The ERICA tool. J. Environ. Radioact. 99, 1371–1383.

Copplestone, D., Bielby, S., Jones, S.R., et al., 2001. Impact Assessment of Ionising Radiation on Wildlife. R&D Publication 128. Environment Agency, Bristol.

Copplestone, D., Wood, M.D., Bielby, S., et al., 2003. Habitat Regulations for Stage 3 Assessments: Radioactive Substances Authorisations. R&D Technical Report P3-101/SP1a. Environment Agency, Bristol.

Hoppeler, H., Weibel, E.R. (Eds.), 2005. Scaling functions to body size: theories and facts. J. Exp. Biol. 208, 1573–1574.

Hosseini, A., Thørring, H., Brown, J.E., et al., 2008. Transfer of radionuclides in aquatic ecosystems – default concentration ratios for aquatic biota in the ERICA tool assessment. J. Environ. Radioact. 99, 1408–1429.

IAEA, 2004. Sediment Distribution Coefficients and Concentration Factors for Biota in the Marine Environment. IAEA Technical Report Series No. 422. International Atomic Energy Agency, Vienna.

IAEA, 2010. Handbook of Parameter Values for the Prediction of Radionuclide Transfer in Terrestrial and Freshwater Environments. IAEA Technical Report Series No. 472. International Atomic Energy Agency, Vienna.

ICRP, 1975. Reference Man: Anatomical, Physiological and Metabolic Characteristics. ICRP Publication 23. Pergamon Press, Oxford.

ICRP, 1988. Limits of intakes of radionuclides by workers: part 4. ICRP Publication 30. Ann. ICRP 19(4), p. 163.

ICRP, 2008. Environmental protection: the concept and use of reference animals and plants. ICRP Publication 108. Ann. ICRP 38(4–6).

Larsson, C.M., 2008. An overview of the ERICA integrated approach to the assessment and management of environmental risks from ionising contaminants. J. Environ. Radioact. 99, 1364–1370.

MacDonald, C.R., 1996. Ingestion Rates and Radionuclide Transfer in Birds and Mammals on the Canadian Shield. Report TR-722 COG-95-551. Atomic Energy of Canada Ltd., Ontario.

Smith, J.T., 2006. Modelling the dispersion of radionuclides following short duration releases to rivers: part 2. Uptake by fish. Sci. Total Environ. 368, 502–518.

Thomann, R.V., 1981. Equilibrium model of fate of microcontaminants in diverse aquatic food-chains. Can. J. Fish. Aquat. Sci. 38, 280–296.

USDOE, 2002. A Graded Approach for Evaluating Radiation Doses to Aquatic and Terrestrial Biota. Technical Standard DOE-STD-1153-2002. United States Department of the Environment, Washington, DC. Available from: <http://homer.ornl.gov/sesa/environment/bdac/manual.html> (accessed August 2011).

USDOE, 2004. RESRAD-BIOTA: a Tool for Implementing a Graded Approach to Biota Dose Evaluation. ISCORS Technical Report 2004-02 DOE/EH-0676. United States Department of the Environment, Washington, DC. Available from: <http://web.ead.anl.gov/resrad/RESRAD_Family/> (accessed August 2011).

Vives i Batlle, J., Wilson, R.C., McDonald, P., 2007. Allometric methodology for the calculation of biokinetic parameters for marine biota. Sci. Total Environ. 388, 256–269.

Vives i Batlle, J., Wilson, R.C., Watts, S.J., et al., 2008. Dynamic model for the assessment of radiological exposure to marine biota. J. Environ. Radioact. 99, 1711–1730.

West, G.B., Brown, J.H., Enquist, B.J., 2000. Scaling in biology: patterns and processes, causes and consequences. In: Brown, J.H., West, G.B. (Eds.), Scaling in Biology. Oxford University Press, Oxford, pp. 87–112.

Wood, M.D., Marshall, W.A., Beresford, N.A., et al., 2008. Application of the ERICA integrated approach to the Drigg coastal sand dunes. J. Environ. Radioact. 99, 1484–1495.

Yankovich, T.L., Vives i Batlle, J., Vives-Lynch, S., et al., 2010. International model validation exercise on radionuclide transfer and doses to freshwater biota. J. Radiol. Prot. 30, 299–340.

3. DERIVATION OF CONCENTRATION RATIOS FOR REFERENCE ANIMALS AND PLANTS

3.1. Collation of data

(54) An online database entitled the 'Wildlife Transfer Database' (http://www.wildlifetransferdatabase.org) was specifically developed to provide parameter values for use in environmental radiological impact assessments to estimate the transfer of radioactivity to non-human biota (i.e. wildlife). The database was initiated to aid: (i) the derivation of CR values for the Commission's list of RAPs as defined at the taxonomic level of family; and (ii) the International Atomic Energy Agency in the production of a handbook of generic wildlife transfer parameters (IAEA, in preparation) for use during environmental assessments. In this way, both organisations have drawn upon the same primary source data in the process of deriving transfer parameters, but there is a distinct difference between the intended use and derivation of the CR values in the respective documents. The database was compiled in collaboration with the International Union of Radioecologists, and it is intended that the database will provide a continuing and evolving source of information on CR values to those conducting assessments and developing/maintaining assessment models. Thus, any new data addressing the data gaps for the RAP CR values can be entered into the online database for use in the future.

(55) The Wildlife Transfer Database was set up with the primary objective of collating data on whole-body CRs, but data can be entered in relation to activity concentrations in organisms and habitat media (e.g. soil or water), along with information concerning the details of the study. Emphasis has been placed on collating empirically derived data from field studies, although information from laboratory experiments has also been included (although no data for RAP species from laboratory studies have been used to derive the values presented here). The location and date for sampled data can be recorded. This information shows that there is reasonable global coverage, although with a bias to those regions where studies on radionuclide transfer have been conducted: Europe, Australia, Japan, and North America. When entering data on media, the operator is prompted to enter information on the bulk activity concentration as opposed to particular phases or (bioavailable) fractions.

(56) The Wildlife Transfer Database incorporates the ERICA transfer databases (Beresford et al., 2008; Hosseini et al., 2008) discussed in the previous section, and also significant additional data contributed by numerous organisations and individuals largely under the auspices of the IAEA (in preparation). All data were quality controlled, by checking against various quality assurance criteria such as numerical consistency and against duplication, before being accepted for the derivation of RAP CR values.

3.2. Categorisation of Reference Animals and Plants

(57) The Wildlife Transfer Database is structured in terms of broad habitats (e.g. terrestrial, marine, and freshwater) and generic wildlife groups which, although not always strictly based on accepted taxonomical classifications, have been selected to be representative of the major types of organisms. Such wildlife groups have also been designed to be generally compatible with the broad categories used in assessment tools, and to represent potential organisms of interest worldwide (IAEA, in preparation).

(58) As discussed above, the Commission has generalised the RAPs to the taxonomic level of family; consequently, this level of taxonomic classification has been used to identify species for which transfer data are available from the published literature and that have been collated within the Wildlife Transfer Database. The family level specified by the Commission is presented in Table 3.1 for each RAP, along with the ecosystem in which that RAP, or its respective life stage, is usually found. A full description of the individual RAPs is given in ICRP (2008).

(59) The relationship between the generic wildlife groups and the corresponding RAPs in the database is shown in Table 3.2. The wildlife group 'subcategory' has also been included, where appropriate, as this reflects the hierarchical categorisation within the database, and is subsequently used in the derivation of surrogate CR values from the broader wildlife groups when data for an RAP–element combination were unavailable (see Section 3.3.2). The database records whether the data being entered are for the adult or a life stage of a particular RAP.

(60) Entered data can also be grouped by organ/tissue type for at least some of the wildlife groups. Although the focus of this work has been the derivation of whole-body CRs, this organisation of the database allows relevant data on transfer to various organs/body parts to be extracted. The issue of heterogeneous distributions of radionuclides within the bodies of animals in terms of implications for exposure has been recognised by the Commission (ICRP, 2008), and is explored further in Section 4.

3.3. Data manipulation and derivation of concentration ratios

(61) The main objective of this report is to derive CR values that are based, as far as possible, upon summarised statistical information from empirical data sets for RAPs with the values primarily derived from field studies. In cases where this was not possible, the aim was to provide surrogate values, the selection of which could be reasonably justified from an understanding of the transfer processes involved, and in all cases to clearly document the provenance of the CR values.

3.3.1. Deriving summarised statistical information for concentration ratios from empirical data sets

(62) The collated data encompassed a wide range of radionuclide–organism (and, in some cases, stable element–organism) combinations, often from different studies

Table 3.1. ICRP Reference Animals and Plants, their life stages, and specified taxonomic families as identified by ICRP (2008). The table lists the species for which data are available within the family groups.

Reference Animal or Plant	Ecosystem	Family	Species for which data are available
Deer	Terrestrial	*Cervidae*	*Alces alces*, *Capreolus capreolus*, *Cervus elaphus*, *Odocoileus hemionus*, *O. virginiannus*
Calf	Terrestrial		
Adult deer	Terrestrial		
Rat	Terrestrial	*Muridae*	*Hydromys chrysogaster*, *Apodemus flavicollis*, *A. sylvaticus*, *Mus domesticus*, *Rattus rattus*
Duck	Terrestrial, freshwater	*Anatidae*	*(Order) Anseres*, *Mergus merganser*, *Anas crecca*, *A. penelope*, *A. platyrhynchos*
Duck egg	Terrestrial		
Adult duck	Terrestrial, freshwater		
Frog	Terrestrial, freshwater	*Ranidae*	*Rana arvalis*, *R. catesbeiana*, *R. clamitans*, *R. esculenta*, *R. pipiens*, *R. temporaria*, *R. terrestris*
Frog egg	Freshwater		
Frog mass of spawn	Freshwater		
Tadpole	Freshwater		
Adult frog	Terrestrial, freshwater		
Trout	Freshwater	*Salmonidae*	*Coregonus albula*, *C. clupeaformis*, *C. hoyi*, *C. lavaretus*, *C. peled*, *Oncorhynchus kisutch*, *O. mykiss*, *O. tschawytscha*, *Prosopium cylindraceum*, *Salmo trutta*, *Salvelinus alpinus*, *S. fontinalis*, *S. namaycush*, *S. siscowet*, *Stenodus leucichthys*
Trout egg	Freshwater		
Adult trout	Freshwater		
Flatfish	Marine	*Pleuronectidae*	*Glyptocephalus stelleri*, *Hippoglossoides dubius*, *Hippoglossus hippoglossus*, *Kareius bicoloratus*, *Limanda herzensteini*, *L. schlencki*, *L. yolohamae*, *Microstomus ache*, *Paralichthys olivaceus*, *Pleuronectes flesus*, *P. platessa*, *Reinhardtius hippoglossoides*, *Synaptura marginata*
Flatfish egg	Marine		
Adult flatfish	Marine		
Bee	Terrestrial	*Apidea*	
Bee colony	Terrestrial		
Adult bee	Terrestrial		
Crab	Marine	*Cancridae*	*Cancer pagarus*
Crab larvae	Marine		
Crab egg mass	Marine		
Adult crab	Marine		
Earthworm	Terrestrial	*Lumbricidae*	*Aporrectodea caliginosa*, *Dendrobaena octaedra*, *Eisenia andrei*, *E. foetida*, *E. nordenskioldi*, *Lumbricus rubellus*, *Lumbricus* spp., *L. terrestris*
Earthworm egg	Terrestrial		
Adult earthworm	Terrestrial		
Pine Tree	Terrestrial	*Pinaceae*	*Abies amabalis*, *Larix decidua*, *L. occidentalis*, *Picea abies*, *Pinus banksiana*, *P. contorta*, *P. strobus*, *P. sylvestris*, *P. taeda*, *Pseudotsuga menziesii*
Wild Grass	Terrestrial	*Poaceae*	*Agropyron cristatum*, *A. dasystachyum*, *Agrostis stolonifera*, *A. tenuis*, *Alopecurus* spp., *Ammophila arenaria*, *Arrhenatherum elatius*, *Bromus arvensis*, *B. inermis*, *B. tectorum*, *Calamagrostis rubescens*, *C. epigeios*, *Cynodon nlemfuensis*, *Dactylis glomerata*, *Deschampsia alpine*, *D. caespitosa*, *D. flexuosa*, *Echinochloa polystachya*, *Erianthus arundinaceum*, *Festuca pratensis*, *F. rubra*, *Hemarthria altissima*, *Holcus mollis*, *Hordeum jubatum*, *Lolium perenne*, *Molinia caerulea*, *Nardus stricta*, *Pennisetum purpureum*, *Phleum pratense*, *Poa pratense*, *Psathyrostachys juncea*, *Puccinellia nuttalliana*, *Sporooulus airoides*, *Stipa viridula*, *Trisetum spicatum*, *Typha latifolia*
Meristem	Terrestrial		
Grass spike	Terrestrial		
Brown Seaweed	Marine	*Fucaceae*	*Fucus disticus*, *F. evanescenes*, *F. inflatus*, *F. serratus*, *F. spiralis*, *F. vesiculosus*

Table 3.2. Wildlife groups (broad group and, where appropriate, its subcategory) and the corresponding Reference Animals and Plants (identified in parentheses under the wildlife group within which they fit).

Freshwater	Marine	Terrestrial
Amphibian (Frog)	Fish *Fish – Benthic Feeding (Flatfish)	Amphibian (Frog)
		Bird (Duck)
Bird (Duck)	Crustacean *Crustacean – Large (Crab)	Arthropod (Bee)
Fish *Fish – Piscivorous (Trout)	Macro-algae (Brown Seaweed)	Grasses and herbs (Wild Grass)
		Mammal (Rat)
		Mammal *Mammal – Herbivorous (Deer)
		Annelid (Earthworm)
		Tree (Pine Tree)

* Subcategory.

with varying sample sizes. Empirical data were not always available in an internally consistent format, and therefore a number of data manipulations were applied. The main conversions (preferentially using information supplied in the source, or associated, references) performed on the data were as follows.

- Where data were presented in the original publication as an activity per unit ash or dry weight, a conversion was required to transform the data to activity per unit fresh weight. If the conversion factors needed were not given in the original publication, an appropriate factor was used from Beresford et al. (2008) or Hosseini et al. (2008).
- Where data were presented in the original publication as an activity concentration for a specific body part or organ, conversion factors were required to transform the data to an activity concentration for the whole body. This data manipulation requires data on total organism live-weight, the component tissues and organs, and the distribution of radionuclides within the body. If appropriate conversion factors could not be derived from the original publication, they were taken from Beresford et al. (2008), Hosseini et al. (2008), or Yankovich et al. (2010a).
- For terrestrial organisms, if transfer data were related to radionuclide deposition (i.e. Bq/m^2 soil rather than Bq/kg), a soil bulk density of 1400 kg/m^3 and a sampling depth of 10 cm were assumed if the source publication lacked the information required to convert the soil activities (Beresford et al., 2008).

(63) There are some uncertainties associated with the data because of lack of information in the original publications, and assumptions have been necessary. For instance, some CR values for aquatic systems may have been estimated using unfiltered water but this was often not specified. Similarly, soil sampling depths were often not given. Such uncertainties will undoubtedly have the effect of introducing greater variability to the data collated than would have existed had these factors been standardised. Furthermore, whilst the CR is assumed to represent an equilibrium transfer value, it is likely that some of the values within the databases were not derived under true equilibrium conditions. To mitigate this problem to some degree, data for terrestrial ecosystems that were collected during the period of above-ground nuclear weapons testing fallout, assumed to be before 1970, or the year of the Chernobyl accident (1986) were not used to derive transfer parameter values for radionuclides of Cs, Pu, Sr, and Am to avoid any effects of surface contamination of vegetation. A full discussion of these issues when deriving CR values for wild species is given by Beresford et al. (2004).

(64) A lack of information in some source publications resulted in some assumptions and compromises having to be made to derive the weighted mean CR values. These were: (i) a sample number of one was assumed if information was not given; (ii) if a measure of error (e.g. standard deviation or standard error) was reported and it was apparent that multiple samples had been collected but no sample number was given, the sample number was assumed to be three; (iii) if a measure of error was only reported for media or biota activity concentrations, this was carried through (proportionally) to give a standard deviation estimate on the calculated CR values; and (iv) a sample number of two was assumed if a minimum and maximum were reported with no details of sample number. For organism–radionuclide combinations with many previously collated (and quality checked) values, new references that did not give all the required information were rejected. Potentially, the summarised value may be skewed in cases where data have been reported in the literature based on a large number of measurements but for which no information on this number was given. However, the assumptions applied were considered to be the least biased (i.e. by avoiding the 'reading' of additional information into a particular case where such information is non-existent), and a logically consistent way of weighting the collated data whilst drawing on all available actual and ancillary information.

(65) CR values from the database for RAPs have been extracted and compiled (see Section 4). A combined weighted (or, for later application, arithmetic) mean (M) and an overall standard deviation value for each CR value was produced using the approach described by Hosseini et al. (2008). It was assumed that the combined variance was comprised of two parts: one describing the variations within studies, and the other expressing the variations between studies. Hence, the total/combined variance can be defined as:

$$V_{combined} = \frac{\left(\sum_{i}(n_i - 1)E_i\right) + \left(\sum_{i} n_i CR_i^2 - NM^2\right)}{N - 1} \tag{3.1}$$

$$N = \sum_i n_i \quad \text{and} \quad M = \frac{\sum_i n_i CR_i}{N}$$

where n_i is the number of observations in study i, CR_i is the mean CR value associated with that study, E_i is the reported measure of error in study i [this can be variance ($E_i = V_i$), standard deviation ($E_i = (Sd)_i^2$), or standard error ($E_i = n_i(Se)_i^2$)], N is the total number of observations in all studies, and M (assumed = M_A as used below) defines the weighted mean composed of means associated with all the considered studies.

(66) The geometric mean, M_G, and geometric standard deviation, σ_G, were approximated using the following equations:

$$M_G = \exp\left(-0.5 \ln\left(\frac{\sigma_A^2 + M_A^2}{M_A^4}\right)\right) \tag{3.2}$$

where σ_A is the (arithmetic) standard deviation of the CR, and M_A is the arithmetic mean of the CR.

$$\sigma_G = \exp\left(\sqrt{\ln\left(\frac{\sigma_A^2 + M_A^2}{M_A^2}\right)}\right) \tag{3.3}$$

where σ_A is the (arithmetic) standard deviation of the CR, and M_A is the arithmetic mean of the CR.

(67) Both the geometric and arithmetic means and standard deviations are presented in this report. In a more general sense, when data sets are large and it is possible to demonstrate statistically that the data are log-normally distributed, the geometric mean provides the most suitable indicator of central tendency and, in conjunction with the geometric standard deviation, characterises the data set most appropriately (Williams et al., 1992). For this work, the observation that many types of radio ecological data tend to exhibit log-normal distributions (Oughton et al., 2008) has led to the assumption that the geometric mean represents the CR values most appropriately for data sets where the sample number exceeds two. For smaller data sets, the arithmetic mean of the CR is selected to be the representative value.

(68) Summarised statistical information derived from empirical CR values that have been collated for species within the Commission's RAP definitions are presented in Annex A.

(69) ^{3}H and ^{14}C CR values in the terrestrial ecosystems were derived using a specific activity approach as described by Galeriu et al. (2003). The approach used for ^{3}H considered both tritiated water and organically bound tritium. For freshwater and marine ecosystems, a simple specific activity approach was applied in numerous cases based on tritiated water alone (IAEA, 2004; Yankovich et al., 2008). The CRs for C were derived through reference to generic reviews (Hosseini et al., 2008).

3.3.2. Deriving surrogate CR values via data-gap-filling methods

(70) As the aim of this report is to provide CR values for all element combinations, a set of rules was considered to facilitate the derivation of surrogate values in cases where limited or no empirical data were available. This also provided a systematic process for documenting how baseline values have been derived in all cases when data were unavailable.

(71) The options used were as follows.

- Use an available CR value for the generic wildlife group 'subcategory' (essentially a subset of the generic wildlife group based on feeding habits or size, e.g. the subcategory 'Crustacean – Large' is part of the generic group 'Crustacean', and the subcategory 'Fish – Benthic Feeding' is part of the generic group 'Fish') within which the RAP fits (e.g. assume the generic Grasses CR value for Reference Wild Grass, or the Mammal – Herbivorous CR value for Reference Deer) (Table 3.2).
- Use an available CR value for the generic wildlife group 'broad group' within which the RAP fits for the radionuclide under assessment (Table 3.2) (e.g. use the generic Fish CR value for Reference Flatfish).
- Use a CR value for a related generic wildlife group 'subcategory' or 'generic group' (e.g. use the Shrub CR for Reference Pine Tree).
- Use an available CR value for the given RAP for an element of similar biogeochemistry (e.g. use the Reference Pine Tree Am CR value for Cm).
- Use an available CR value for biogeochemically similar elements for the generic wildlife group (generic group or subcategory) within which the RAP fits (e.g. use the Fish Am CR value for Cm in Reference Flatfish).
- Use allometric relationships, or other modelling approaches, to derive appropriate CRs.
- Expert judgement of CR data within that ecosystem for the radionuclide under assessment, which might include, for example, the use of data from general reviews on this subject.
- In the case of the marine ecosystem, use CR data from the estuarine ecosystem.

Where the options yielded poorly supported (e.g. in terms of number of underpinning data) alternative values, the selection of the most conservative value was often made. In all cases, the reasoning underpinning the selection of values is recorded transparently.

(72) Although the first and second methods listed above may be considered the preferred options, this may not always be true. For example, very few data may be available for these options, but numerous data may be available for subsequent options. Thus, an element of subjective judgement was sometimes required in deriving some values; where this has occurred, this has been documented.

(73) Although the methods to fill data gaps adopted here have not been validated explicitly, in cases where similar approaches have been applied, the predictions of activity concentrations in biota derived from media concentrations are reasonably consistent with directly measured values. The international intercomparisons described by Yankovich et al. (2010b) and Beresford et al. (2010) for

freshwater and terrestrial scenarios, respectively, provide evidence for this contention. In many cases, this is not surprising because differences in transfer between a particular RAP defined at the family level and the broader generic wildlife group are likely to be small. A comparison between the terrestrial generic wildlife group and the terrestrial RAP CR values show a few (out of 90) values which differ by more than an order of magnitude; the majority differ by less than a factor of four. In the aquatic environments, there are no obvious physiological reasons to expect that the CRs for Reference Trout in the freshwater environment and Reference Flatfish in the marine environment should differ dramatically from CRs for their respective broader wildlife groups in the guise of generic freshwater and marine fish, respectively, which have often been used in deriving surrogate values. There are, of course, some exceptions to the expectation that taxonomically related organisms will express similar degrees of transfer, as exemplified by the case of elevated ^{99}Tc uptake in some groups of crustaceans compared with others (Brown et al., 1999).

(74) If information about the activity concentrations of radionuclides in components of an animal's diet can be acquired, along with parameters for ingestion rate [that can be allometrically derived as shown by Nagy (2001)], assimilation efficiencies, and biological half-lives [many of which can also be derived using allometric approaches, as demonstrated by Higley et al. (2003)], equilibrium CR values can be estimated. Nonetheless, there are certain caveats when applying this approach: (i) the derivation methods are likely to rely upon CR approaches for the base of a food chain (i.e. they may be no better at accounting for variability resulting from soil type, water chemistry, etc.); and (ii) it is not clear whether this method can provide predictions for many radionuclides with any more confidence than the CR approach.

(75) Summarised statistical information derived from empirical data specifically for CR values for generic wildlife groups that encompass RAPs are reported elsewhere (IAEA, in preparation). Surrogate CR data with a detailed description of how values have been derived are presented in Annex B. Annexes A and B provide coverage of CR values for all element–RAP combinations.

(76) When the data underpinning a RAP CR value are limited, there may be some justification in selecting a CR value from the more generic wildlife group, as in the IAEA report (IAEA, in preparation) where the data set is more extensive. A larger data set arguably provides a more realistic indication of the variance, and provides more extensive information for deriving a suitable best estimate. However, this approach has not been used in this report because of the self-imposed requirement to provide CR values solely for the RAPs, as defined by the Commission. Consequently, this report has reported CR values explicitly for the 12 families of organisms where they exist, irrespective of how few data the values are based upon.

3.3.3. Concentration ratio values

(77) CR values for the RAPs (see Section 4) have been extracted from the summary statistics and derived values presented in Annexes A and B, respectively. The

CR values are essentially a single value representation for each element–RAP combination with their provenance described. Where empirical data exist explicitly for RAPs, the CR value is based on the geometric mean reported in Annex A. The underlying transfer data sets are generally assumed to follow log-normal distributions, and the geometric mean provides the most suitable measure of central tendency in such circumstances. The arithmetic mean value is used to derive the CR value when $n \leqslant 2$. In cases where no empirical data exist specifically for the RAP, the derived values presented in Annex B have been used to provide the CR value. For these derived data, the arithmetic mean generic wildlife CR value is used where $n \leqslant 2$, and the geometric mean for the generic wildlife group is used where $n > 2$.

(78) To date, no attempt has been made to derive CR values for the various life stages of RAPs, for reasons discussed in Section 4.

3.4. References

Beresford, N.A., Broadley, M.R., Howard, B.J., et al., 2004. Estimating radionuclide transfer to wild species – data requirements and availability for terrestrial ecosystems. J. Radiol. Prot. 24, A89–A103.

Beresford, N.A., Barnett, C.L., Howard, B.J., et al., 2008. Derivation of transfer parameters for use within the ERICA tool and the default concentration ratios for terrestrial biota. J. Environ. Radioact. 99, 1393–1407.

Beresford, N.A., Barnett, C.L., Brown, J., et al., 2010. Predicting the radiation exposure of terrestrial wildlife in the Chernobyl exclusion zone: an international comparison of approaches. J. Radiol. Prot. 30, 341–373.

Brown, J.E., Kostad, A.K., Bringot, A.L., et al., 1999. Levels of ^{99}Tc in biota and seawater samples from Norwegian coastal waters and adjacent sea. Mar. Pollut. Bull. 38, 560–571.

Galeriu, D., Beresford, N.A., Melintescu, A., et al., 2003. Predicting tritium and radiocarbon in wild animals. International Conference on the Protection of the Environment from the Effects of Ionizing Radiation, 6–10 October 2003, Stockholm, IAEA-CN-109, International Atomic Energy Agency, Vienna, pp. 186–189.

Higley, K.A., Domotor, S.L., Antonio, E.J., 2003. A kinetic–allometric approach to predicting tissue radionuclide concentrations for biota. J. Environ. Radioact. 66, 61–74.

Hosseini, A., Thørring, H., Brown, J.E., et al., 2008. Transfer of radionuclides in aquatic ecosystems – default concentration ratios for aquatic biota in the ERICA tool assessment. J. Environ. Radioact. 99, 1408–1429.

IAEA, 2004. Sediment Distribution Coefficients and Concentration Factors for Biota in the Marine Environment. IAEA Technical Reports Series No. 422. International Atomic Energy Agency, Vienna.

IAEA, in preparation. Handbook of Parameter Values for the Prediction of Radionuclide Transfer to Wildlife. IAEA Technical Report Series. International Atomic Energy Agency, Vienna.

ICRP, 2008. Environmental protection: the concept and use of reference animals and plants. Publication 108. Ann. ICRP 38(4–6).

Nagy, K.A., 2001. Food requirements of wild animals: predictive equations for free-living mammals, reptiles and birds. Nutr. Abs. Rev. Ser. B 71, 21–31.

Oughton, D.H., Agüero, A., Avila, R., et al., 2008. Addressing uncertainties in the ERICA integrated approach. J. Environ. Radioact. 99, 1384–1392.

Williams, A.C., Cornwell, P.A., Barry, B.W., 1992. On the non-Gaussian distribution of human skin permeabilities. Int. J. Pharm. 86, 69–77.

Yankovich, T.L., Sharp, K.J., Benz, M.L., Carr, J., Killey, R.W.D., 2008. Carbon-14 specific activity model validation for biota in wetland environments. Proceedings of the 2007 ANS Topical Meeting on Decommissioning, Decontamination, and Reutilization, 16–19 September 2007, Chattanooga, TN, American Nuclear Society, Illinois, p. 336.

Yankovich, T.L., Beresford, N.A., Wood, M.D., et al., 2010a. Whole body to tissue-specific concentration ratios for use in biota dose assessments for animals. Radiat. Environ. Biophys. 49, 549–565.

Yankovich, T.L., Vives i Batlle, J., Vives-Lynch, S., et al., 2010b. An international model validation exercise on radionuclide transfer and doses to freshwater biota. J. Radiol. Prot. 30, 299–340.

4. CONCENTRATION RATIOS FOR REFERENCE ANIMALS AND PLANTS

4.1. Applicability of concentration ratios for Reference Animals and Plants

(79) In considering how RAPs are exposed to radioactive substances, it is necessary to explore how internal exposures (as indicated by whole-body activity concentrations) are related to the radionuclide content of the surrounding environment. When using CR values, it is assumed that these two quantities are correlated. However, as noted in Section 2, this may not always be true. At a generic level, equilibrium conditions are required when critically applying CRs. However, in many instances, the concentrations of radionuclides in environmental media may fluctuate. This temporal variability is influenced by various environmental factors affecting the input to, and losses from, media compartments, such as water residence times in freshwater and marine systems, and composition and structural-dependent leaching rates for soils in terrestrial systems. Furthermore, equilibrium between the different RAPs and environmental media will be dependent upon a number of factors (e.g. biological half-life, lifespan) which are environment, radionuclide, and RAP specific.

(80) Many factors may modify uptake and therefore reduce the applicability of generic values to specific cases. Studies have demonstrated that uptake may be competitively inhibited by the presence of 'analogous' ions (Shaw and Bell, 1991), which means that transfer for any given radionuclide might be expected to deviate in relation to variations in the concentrations of stable element analogues.

(81) For terrestrial systems, soils are known to vary widely in terms of their lithology and chemical composition. Soil type can affect the bioavailability of elements and their potential for transfer through terrestrial food chains (IAEA, 2010). Soil types have been used to categorise the degree of transfer to various crops in tropical and subtropical environments (Velasco et al., 2009). Categorising CR values by soil type has not been attempted in the present collation exercise due to a paucity of data that would allow this to be achieved convincingly. Nonetheless, the effect of soil type on transfer of radionuclides to RAPs might be worthy of consideration in future iterations of statistical analyses, once larger data sets become available.

(82) The denominator for terrestrial CR values is based upon total soil activity concentrations, whereas the uptake to plants and subsequent transfer to herbivores and predators is likely to reflect the bioavailable fraction of radionuclides which may or may not reflect the total soil activity concentration. There is a case to answer, therefore, that the approach as a predictive tool has certain limitations, although for practical reasons, most notably from the consideration that determinations of activity concentrations are normally made for bulk soils, it was considered the only viable approach.

(83) The chemical composition of a water source is known to affect the uptake of many radionuclides. For example, Kolehmainen et al. (1966), having classified lakes according to numerous physical, chemical, and biological properties, determined that the highest levels of ^{137}Cs were observed in fish from oligotrophic lakes. In particular, the importance of K^{+} ions in affecting ^{137}Cs uptake by fish is well documented (Blaylock, 1982; Fleishman et al., 1994).

(84) The chemical composition of marine water tends to be much less variable than that observed in freshwater environments, which may reduce one source of variance in CR values. Nonetheless, the relationship between the concentration of an element or radionuclide in a living organism and the ambient seawater is dynamic. Rates of both uptake and excretion are known to be affected by body size, rate of change of body size, temperature, light (in the case of algae), salinity etc. (IAEA, 2004). The bioavailability of certain metals in seawater may differ greatly with oxidation state [cf. Pu (III) vs Pu (V)]. Seasonal variation in metal CRs in marine organisms may also be important under certain conditions (IAEA, 2004).

(85) When exposure becomes elevated beyond the normal background (e.g. through mining activity), an ecological succession may result with a transition to plant species that exhibit distinctly different transfer characteristics including the evolution of hyperaccumulation of heavy metals by some species of plants (e.g. Baker et al., 1988). With this consideration in mind, the summarised data were inspected for the presence of potential hyperaccumulating species, and as a consequence, some values for Se and U were not used in the derivation of summary values (see IAEA, in preparation).

(86) Variability can also be introduced to the underlying CR data set as a result of non-standardised methods for deriving whole-body CR data, such as the inclusion of some data for which the gut was not purged prior to analysis (i.e. the data are for whole-body concentrations including the stomach contents). This may be true, especially for studies of smaller organisms where it is more difficult to separate the gut prior to analysis. The activity concentration of gut contents can vary substantially from the body as a whole, and may result in quite different whole-body CR values from those derived without the gut contents. Furthermore, substantial variability may be introduced to the CR derivation through non-standardised selection of environmental media samples, especially for migratory organisms, those animals with a large home range, or where lifecycle changes result in changes of habitat.

(87) Notwithstanding these limitations, CR values have been widely applied, as noted in Section 3, and have either been derived from field data or from laboratory experiments. Field data are dependent on factors such as biological half-lives, physical half-lives, ecological characteristics (e.g. water chemistry), and source term, but these factors are generally more likely to resemble those under which the CRs are applied within an assessment. Hence the focus has been on in-situ empirical data collation.

(88) For some organisms, many elements (and thus their radioisotopes or radionuclides of analogous elements) are under some form of homeostatic control that regulates their concentrations internally, irrespective of fluctuations in their intake (e.g. via food or water), and are thus not affected by the ambient concentration levels in the environment. For example, potassium is controlled homeostatically in higher animals, and the resulting concentration of potassium is effectively constant (e.g. Koulikov and Meili, 2003). Given the constant ratio between ^{40}K and stable K, activity concentrations of ^{40}K in the body will also be constant.

(89) ^{40}K is also an important component of the natural background, and as such is likely to be characterised by direct measurements. For this reason, CRs have not

been considered further for this particular radionuclide. Natural background exposures to RAPs from naturally occurring radionuclides including ^{40}K have been published elsewhere (Beresford et al., 2008b; Hosseini et al., 2010).

(90) Stable element data have often been used in the derivation of aquatic CRs, and these may well be better representative values of steady-state conditions. Elemental concentrations in seawater for many, but not all, elements are reasonably constant (Millero, 1996), and hence the application of stable data to derive CR values is relatively well founded. The situation is, however, very different for freshwater in the sense that these ecosystems are often associated with highly variable dissolved element concentrations ranging from those characteristic of oligatrophic through to eutrophic water bodies. The (often inverse) correlation between radionuclide activity concentrations in organisms and stable analogue concentrations in water, as alluded to above, suggests that freshwater CRs should ideally be categorised according to water chemistry. Lack of ancillary information in the collated data, however, made this impracticable at the time of writing this report for the range of RAPs and elements considered.

(91) The reliance on stable element data in the derivation of CRs in some cases may also limit the applicability of the values to radionuclides in situ. This is particularly true for short-lived radionuclides when the physical half-life is considerably shorter than the biological half-life. Notable examples exist for the cases of transfer data derived from stable phosphorus (in its application to ^{32}P and ^{33}P) and stable sulphur (in its application to ^{35}S).

4.2. Concentration ratio values for terrestrial Reference Animals and Plants and their applicability

(92) CR data for adult terrestrial RAPs are presented in Tables 4.1 and 4.2. These data are based on the detailed tables reported in Annexes A and B, which include full references.

(93) Empirical CR data for terrestrial animals and plants are sporadic. The data coverage for some RAPs (Reference Earthworm, Reference Wild Grass, and Reference Pine Tree) is reasonable, extending to 15 or more of the elements considered, although some of the values reported are based on single measurements in some instances. However, for Reference Rat, Reference Frog, Reference Deer, and Reference Duck, only four to 10 elements were covered. For Reference Bee, no specific data (i.e. for the family *Apidea*) were found. Consequently, for the terrestrial RAPs, there is a reliance on the use of derived (or surrogate) CR values. These surrogate values were generally derived from the generic wildlife groups, but in a few cases, recourse was made to element analogues.

(94) Reference Wild Grass appears to lend itself most readily to the CR approach, because many elements are obtained directly from the medium in which the plant is growing, and thus the link between activity concentrations in the plant tissues and soil might be considered to be clearly evident. Some elements (e.g. C) are incorporated via direct exchange with elements in the ambient atmosphere, and

Table 4.1. Concentration ratio (CR) values (geometric mean or best-estimate-derived value in units of Bq/kg fresh weight per Bq/kg dry weight soil or per Bq/m^3 for C, H, S, and P) for adult terrestrial Reference Animals and Plants – invertebrates and plants.

Element	Bee	Earthworm	Wild Grass	Pine Tree
Ag	7.0E−01(f)	7.0E−01(b)	1.8E+00(a)	1.9E−02(b)
Am	4.0E−02(a)	1.1E+00	1.5E−01	1.7E−02(b)
Ba	3.8E−02(a)	3.8E−02(b)	5.4E−02(a)	1.6E−01
C	4.3E+02(e,b)	4.3E+02(e)	8.9E+02(e)	1.3E+03(e)
Ca	1.0E+01(f)	1.0E+01(f)	2.2E+00(f)	5.0E+00(f)
Cd	1.4E+00(a)	3.6E+00	2.7E+00	3.5E−01(a)
Ce	3.7E−04(b)	3.7E−04	3.6E−03(a)	3.3E−03
Cf	4.0E−02(d)	1.1E+00(a)	3.3E−02(c)	4.3E−02(d)
Cl	2.8E−01(a)	1.7E−01	4.9E+01	1.1E+00
Cm	1.4E−01(d)	1.1E+00(c)	5.0E−04(a)	9.4E−03(b)
Co	4.7E−03(a)	4.7E−03(f)	3.9E−03(a)	1.4E−03
Cr	5.0E−03(f)	5.0E−03(f)	5.8E−03(a)	3.8E−03
Cs	4.7E−03(a)	4.8E−02	8.6E−01	7.5E−02
Eu	7.9E−04(b)	7.9E−04	3.6E−03(a)	2.1E−03
H	1.5E+02(e)	1.5E+02(e)	1.5E+02(e)	1.5E+02(e)
I	2.8E−01(a)	1.4E−01	5.3E−02(a)	5.3E−02(b)
Ir	4.1E−03(d)	4.1E−03(d,f)	4.0E−02(c,f)	3.2E−01(d,f)
La	3.7E−04(b)	3.7E−04(c)	6.0E−03(a)	3.1E−03
Mn	4.4E−02(b)	1.3E−02	1.6E−01(f)	2.4E−02(a)
Nb	5.1E−04(b)	5.1E−04	5.0E−03(f)	5.0E−03(f)
Ni	8.6E−03(a)	2.3E−02	1.8E−01	1.8E−02(f)
Np	4.0E−02(d)	1.1E+00(c)	1.5E−02(f)	4.3E−02(d)
P	4.3E+02(d,b)	4.3E+02(d,e)	8.9E+02(d,f)	1.3E+03(d,f)
Pa	4.0E−02(d)	1.1E+00(c)	3.3E−02(c)	4.3E−02(d)
Pb	2.6E−01(a)	5.7E−01	7.5E−02	5.3E−02
Po	9.6E−02(b)	9.6E−02	2.3E−01	4.0E−02
Pu	1.6E−02(a)	2.1E−02(a)	3.3E−02	4.3E−02(b)
Ra	2.1E+00(a)	2.1E+00(b)	9.2E−02	6.3E−04
Ru	4.1E−03(a)	4.1E−03(b)	4.0E−02(f)	3.2E−01(b)
S	5.0E+01(f)	5.0E+01(f)	1.5E+02(f)	1.5E+02(f)
Sb	1.8E−01(b)	6.0E−03	4.1E+01	3.2E+00(b)
Se	1.5E+00(b)	1.5E+00	1.3E+00	1.1E+00(b)
Sr	8.4E−02(a)	9.0E−03	1.7E+00	2.0E−01
Tc	3.5E−01(f)	3.5E−01(f)	3.2E+00	8.4E−03(b)
Te	3.8E−02(f)	3.8E−02(f)	2.5E−01(f)	2.5E−01(f)
Th	1.7E−02(d)	8.8E−03(f)	9.5E−02	3.2E−04
U	1.7E−02(a)	8.8E−03	4.3E−02	9.9E−04
Zn	9.7E−01(a)	3.7E+00	2.6E+00	3.5E−02
Zr	5.1E−04(d)	5.1E−04(c)	2.5E−03(f)	7.2E−05(b)

Shaded values were derived using surrogate CR values and according to the code given in brackets.

(a) CR value for the generic wildlife group (i.e. 'generic group' or 'subcategory') within which the Reference Animal or Plant fits for a given element.

(b) CR value derived from a related Reference Animal or Plant or related generic wildlife group for a given element.

(c) CR value for the given Reference Animal or Plant for an element of similar biogeochemistry.

(d) CR value for biogeochemically similar elements for encompassing or related generic wildlife group.

(e) Allometric relationships or other modelling approach.

(f) Expert judgement excluding the approaches explicitly noted above and including data derived from published reviews.

Table 4.2. Concentration ratio (CR) values (geometric mean or best-estimate-derived value in units of Bq/kg fresh weight per Bq/kg dry weight soil or per Bq/m^3 for C, H, S, and P) for adult terrestrial Reference Animals – vertebrates.

Element	Rat	Deer	Duck	Frog
Ag	2.9E−01	2.9E−01(f)	2.9E−01(f)	2.9E−01(f)
Am	3.6E−04	2.1E−03	2.8E−02	1.0E−01
Ba	4.8E−03(a)	4.8E−03(a)	4.8E−03(b)	4.8E−03(b)
C	1.3E+03(e)	1.3E+03(e)	1.3E+03(e)	1.3E+03(e)
Ca	2.0E+00(f)	2.0E+00(f)	2.0E+00(f)	2.0E+00(f)
Cd	7.2E−01(a)	6.7E+00(a)	7.2E−01(b)	1.3E−02
Ce	6.1E−04(e,f)	6.1E−04(e,f)	6.1E−04(e,f)	6.1E−04(e,f)
Cf	1.9E−02(c)	2.1E−03(c)	2.8E−02(c)	1.0E−01(c)
Cl	7.0E+00(e,f)	7.0E+00(e,f)	7.0E+00(e,f)	7.0E+00(e,f)
Cm	1.9E−02(c)	2.1E−03(c)	2.8E−02(c)	1.0E−01(c)
Co	1.8E−01	1.8E−01(a)	1.8E−01(b)	1.8E−01(b)
Cr	2.0E−04(f)	2.0E−04(f)	2.0E−04(f)	2.0E−04(f)
Cs	2.2E−01	1.6E+00	2.2E−01	2.8E−02
Eu	2.0E−03(e,f)	2.0E−03(e,f)	2.0E−03(e,f)	2.0E−03(e,f)
H	1.5E+02(e)	1.5E+02(e)	1.5E+02(e)	1.5E+02(e)
I	4.0E−01(e,f)	4.0E−01(e,f)	4.0E−01(e,f)	4.0E−01(e,f)
Ir	1.2E−01(d,e,f)	1.2E−01(d,e,f)	1.2E−01(d,e,f)	1.2E−01(d,e,f)
La	6.1E−04(d,e,f)	6.1E−04(d,e,f)	6.1E−04(d,e,f)	6.1E−04(d,e,f)
Mn	2.4E−03(a)	2.4E−03(a)	2.4E−03(b)	2.4E−03(b)
Nb	1.9E−01(f)	1.9E−01(f)	1.9E−01(f)	1.9E−01(f)
Ni	7.2E−02(a)	7.2E−02(a)	3.1E−01(b)	3.0E−01(b)
Np	1.9E−02(c)	8.9E−04(c)	2.8E−02(c)	1.0E−01(c)
P	1.3E+03(c,e)	1.3E+03(c,e)	1.3E+03(c,e)	1.3E+03(c,e)
Pa	1.9E−02(c)	8.9E−04(c)	2.8E−02(c)	1.0E−01(c)
Pb	9.6E−03	1.2E−02(a)	2.1E−02(a)	2.6E−03
Po	7.5E−04	2.4E−03(a)	9.6E−03(a)	3.3E−02(b)
Pu	1.9E−02	8.9E−04	1.0E−02	9.3E−03(b)
Ra	4.4E−02	6.1E−03(a)	5.5E−02	1.7E−02(b)
Ru	1.2E−01(e,f)	1.2E−01(e,f)	1.2E−01(e,f)	1.2E−01(e,f)
S	5.0E+01(f)	5.0E+01(f)	5.0E+01(f)	5.0E+01(f)
Sb	6.0E−02(f)	6.0E−02(f)	6.0E−02(f)	6.0E−02(f)
Se	1.0E−02(a)	1.0E−02(a)	1.0E−02(b)	1.0E−02(b)
Sr	2.2E+00	2.1E+00	1.1E−01	1.1E+00
Tc	3.7E−01(e,f)	3.7E−01(e,f)	1.7E−01	3.5E−01(a)
Te	2.1E−01(f)	2.1E−01(f)	2.1E−01(f)	2.1E−01(f)
Th	6.3E−05	1.0E−04(a)	3.8E−04(a)	7.6E−02(b)
U	6.5E−04	3.7E−03(a)	4.9E−04(a)	6.7E−01(b)
Zn	9.2E−02(b)	9.2E−02(b)	9.2E−02(b)	9.2E−02(b)
Zr	1.2E−05(f)	1.2E−05(f)	1.2E−05(f)	1.2E−05(f)

Shaded values were derived using surrogate CR values and according to the code given in brackets.

(a) CR value for the wildlife group (i.e. 'generic group' or 'subcategory') within which the Reference Animal fits for a given element.

(b) CR value derived from a related Reference Animal or related generic wildlife group for a given element.

(c) CR value for the given Reference Animal for an element of similar biogeochemistry.

(d) CR value for biogeochemically similar elements for encompassing or related generic wildlife group.

(e) Allometric relationships or other modelling approach.

(f) Expert judgement excluding the approaches explicitly noted above and including data derived from published reviews.

for atmospheric releases, plant surfaces may be contaminated through the processes of dry and wet deposition (Pröhl, 2009) which may complicate the correlation between soil and plant concentrations depending upon the contamination scenario. With regards to elements where significant atmospheric exchange is occurring (i.e. C, H, P, and S), efforts have been made to address this in the derivation of terrestrial CR values by relating activity concentrations in Reference Wild Grass (and for all other RAP categories for that matter) to those in air. Where aerial discharges have occurred over long time scales, or in cases where long time periods (several months to years) have elapsed following a pulsed or accidental release, CRs should, however, provide a reasonable indication of transfer to Reference Wild Grass and are an accepted approach in human food-chain modelling (e.g. IAEA, 2010). Most data for Reference Wild Grass were collated through direct reference to *Poaceae* data and the generic wildlife group 'Grasses and Herbs', although recourse to the pasture CR value from IAEA (2010) was required in some instances.

(95) The transfer pathways for Reference Pine Tree are similar to those for Reference Wild Grass. However, the fact that trees are long lived means that many elements have the potential to be incorporated within non-living tissue which might affect the CR value. For example, IAEA (2001) showed that the inventory of radiocaesium expressed as a percentage of total phytomass increased in stem wood for pine forests following the Chernobyl accident. This may result in equilibration between soil and plant not occurring, even over protracted time periods. Moreover, the CR values for terrestrial RAPs generally, but for Reference Pine Tree in particular, strongly depend on the soil depth sampled. CR data were related to radionuclide activity concentration in soil with an assumed 10-cm sampling depth (unless the depth was given in the original publication), as described above. In view of rooting depths for trees which normally tend to be in excess of 10 cm, there are concerns regarding whether the most appropriate activity concentration for the denominator in the CR value has been selected for this particular RAP. Many of the data derived for Reference Pine Tree have been extracted from the related wildlife groups 'Tree – Broadleaf' and 'Shrub'.

(96) Reference Earthworm lives in soil and derives nutrition from organic matter in a wide variety of forms, including plant matter (various forms, fresh-decayed), protozoans, rotifers, nematodes, bacteria, fungi, and decomposing remains of other animals. The link between body concentrations of elements in earthworms and soil, at least for those that are not homeostatically controlled, is clear. Published literature may contain earthworm data with and without the gastrointestinal tract and its contents, which will undoubtedly increase the variability in the derived CR values. Nahmani et al. (2007) showed that soil properties can affect the accumulation of metals by earthworms and that this can lead to apparently contradictory results. In some studies, the soil properties appeared to have no or very little influence on accumulation of metals by earthworms, but other studies showed that soil total metal concentration alone was a poor predictor of uptake due to modifying factors such as pH, organic matter content, and clay size particle content (Nahmani et al., 2007). CR data were only available for the phylum *Annelid* in a limited number of

cases, and such values have been used where possible. Nonetheless, a heavy reliance on values for related wildlife groups was required when deriving the full suite of CR values.

(97) Reference Bee spends a large part of its life away from direct contact with soil. During the process of gathering nectar and/or pollen, bees have an indirect route of transfer from soil in the sense that soil provides the source for many radionuclides in the plants that bees habitually visit. Many of the derived data for Reference Bee in this report have been based upon values for the generic wildlife group defined by phylum *Arthropod,* although in some cases, data from related invertebrate wildlife groups were required in filling gaps in transfer information.

(98) Reference Duck, as defined at the level of family *Anatidae*, consists of a number of species that generally undertake annual migrations. Although they may spend up to several months at any single location, the degree of equilibration the bird attains with soil in this time is clearly debatable. Furthermore, ducks spend time on land, on water, and in air. However, most CR values in the literature are often given to soil (or water) taken in the vicinity of where the duck was sampled, and consequently these might not be appropriate for migratory species and should be used with care. There are few direct empirical CR values for Reference Duck. Some additional information becomes available when the search is broadened to include the generic group 'Bird' (CR values for which have been used as a first preference). In numerous cases, CR data for other related generic wildlife groups, such as 'Reptile' and 'Mammal', have been used as surrogates.

(99) The fact that adult frogs spend the majority of their time in both terrestrial and freshwater environments raises questions about which environmental media should be used to estimate body concentrations. In this regard, it would seem sensible to consider transfer from both soil and water, and applying a weighting to the calculations of exposure according to how much time the animal spends in habitats characterised by these media types. Thus, CR data for Reference Frog have been collated for both terrestrial and freshwater ecosystems in this report. Numerous surrogate CR values for Reference Frog have been based on data from mammals, reflecting the general paucity of transfer information for amphibia.

(100) The final categories of terrestrial RAPs are the mammals, Reference Rat and Reference Deer. Both animals derive body burdens primarily through ingestion of food and water and, to a lesser extent, via inhalation (of gases and/or dust) depending on the radionuclide under consideration. Ideally, any CR values presented should have considered the home range of the species in question, and this is particularly true for Reference Deer which may have a large home range. Consequently, spatial averaging should be taken into account when deriving CR values. In practice, the ad-hoc nature of the studies that have been used to populate the underlying Wildlife Transfer Database mean that such considerations have not been applied systematically. Some radionuclide CR values are extremely well characterised, as exemplified by Cs and Sr for *Cervidae*, but in many cases, recourse was made to the various gap-filling approaches, in particular those involving the use of data from the generic wildlife group 'Mammal'.

4.3. Concentration ratio values for freshwater Reference Animals and Plants and their applicability

(101) CR data for adult freshwater Reference Animals are presented in Table 4.3. These data are based on the detailed tables reported in Annexes A and B which include full references.

(102) CR data specifically for freshwater Reference Animals are characterised by a fairly extensive coverage of elements for Reference Trout, but far fewer elements for Reference Frog and none for Reference Duck. The lack of data for Reference Duck may appear surprising as species of duck are often included in human food-chain monitoring programmes. However, such measurements are rarely made in conjunction with analyses of freshwater samples.

(103) The limitations in relation to application of CRs for Reference Duck have been discussed above in relation to soil, and similar points are pertinent with regards to CR values for freshwater. The lack of empirical data on aquatic-based CR values for Reference Duck or the generic wildlife group 'Bird' lead to the use of simple allometric–biokinetic models to derive surrogate values. Assuming a duck of mass 1.26 kg [as specified in ICRP (2008)], fresh matter ingestion rates were derived using the allometric relationships for birds provided by Nagy (2001). Assuming the duck is herbivorous, the CR data for vascular plants can be used from the Wildlife Transfer Database which provides reasonable coverage of the elements of interest. Finally, information on the assimilation efficiency (gut uptake) and biological half-life can be obtained from studies on mammals which, in view of broad similarities in physiology, are believed to provide a reasonable surrogate for birds.

(104) The life cycle of *Ranidae* is complex and home ranges can cover considerable areas (Watson et al., 2003), but the applicability of CRs should be reasonable where an appreciation of these factors is adopted. Many of the CR data for Reference Frog in relation to the freshwater ecosystem have been derived from the CR data for the related, generic wildlife group 'Fish'. This is arguably tenuous in view of the differing physiology and ecology of amphibians compared with fish, but for radionuclides where the ingestion pathway strongly influences body burdens, one might expect, in view of some similarities in diet, there to be a degree of correlation in terms of transfer between these two organism groups. Nonetheless, no analysis (statistical or otherwise) could be performed to establish whether this approach is reasonable because of the lack of data.

(105) The use of a trout rather than a salmon as the reference was deliberate in order to avoid the complication of migratory effects of the salmon from freshwater to the marine environment (ICRP, 2008). This consideration should assist, to some degree, in reducing variability associated with the application of a CR value as prolonged contact between the organism and its surrounding media, conditions conducive to equilibration, can be more safely assumed for non-migratory fish. For Reference Trout, as for other freshwater RAPs, the derivation of a generic CR based on an arbitrary suite of sampling locations, with differing unspecified water chemistries, might not be ideal. Steady-state conditions are unlikely to exist for many

Table 4.3. Concentration ratio (CR) values (geometric mean or best-estimate-derived value in units of Bq/kg fresh weight per Bq/l water) for adult freshwater Reference Animals.

Element	Trout	Frog	Duck
Ag	1.0E+02(f)	1.0E+02(f)	1.0E+02(f)
Am	5.7E+02(a)	5.7E+02(b)	2.7E+02(e)
Ba	8.1E+00	4.3E+01(b)	3.9E+02(d)
C	4.3E+04	7.3E+03(f)	7.3E+03(f)
Ca	3.5E+02	8.5E+02	3.9E+02(b)
Cd	1.9E+02(a)	1.9E+02(b)	1.3E+03(b)
Ce	2.7E+02	6.5E+01(b)	6.3E+02(b)
Cf	2.0E+01(c)	3.8E+01(d)	2.5E+02(c)
Cl	1.3E+02(a)	1.3E+02(b)	8.2E+01(f)
Cm	5.7E+02(d)	5.7E+02(d)	2.7E+02(c,e)
Co	9.5E+01	8.2E+01(b)	4.9E+02(e)
Cr	1.5E+02	6.5E+01	6.8E+00(e)
Cs	2.7E+03	1.6E+03(b)	4.4E+02(e)
Eu	3.0E+01	5.3E+01(b)	5.0E+01(f)
H	1.0E+00(e)	1.0E+00(e)	1.0E+00(e)
I	6.2E+01	2.6E+02(b)	2.2E+02(e)
Ir	2.9E+01(d)	2.9E+01(d)	2.0E+01(f)
La	2.0E+02	6.0E+01(b)	2.4E+02(b)
Mn	3.0E+03	8.6E+02(b)	1.5E+02(b)
Nb	4.9E+02(c)	5.4E+01(d)	2.3E+02(f)
Ni	1.3E+01	1.2E+02(b)	9.5E+02(b)
Np	2.0E+01(c)	3.8E+01(b)	2.5E+02(c,e)
P	7.0E+05	6.4E+05(b)	6.2E+04(f)
Pa	2.0E+01(c)	3.8E+01(b)	2.5E+02(c,e)
Pb	9.1E+01	5.3E+00	5.4E+00(e)
Po	1.6E+02	5.9E+02(b)	1.5E+02(e)
Pu	2.0E+01	3.8E+01(b)	2.5E+02(e)
Ra	3.4E+01	5.5E+01(b)	3.7E+02(b)
Ru	2.9E+01(a)	2.9E+01(b)	2.0E+01(f)
S	8.0E+02(f)	8.0E+02(f)	8.0E+02(f)
Sb	9.8E+00	1.1E+01(b)	2.3E+03(b)
Se	5.9E+03	4.0E+03(b)	1.9E+03(b)
Sr	1.0E+02	1.5E+02(b)	4.1E+03(e)
Tc	7.1E+01(a)	7.1E+01(b)	4.0E+01(f)
Te	2.8E+02(a)	2.8E+02(b)	7.0E+02(f)
Th	9.8E+01(a)	9.8E+01(b)	2.8E+03(e)
U	8.5E+00	9.1E+00(b)	1.3E+01(e)
Zn	1.0E+04	7.3E+02	1.2E+04(e)
Zr	4.9E+02	5.4E+01(b)	1.2E+03(b)

Shaded values were derived using surrogate CR values and according to the code given in brackets.

(a) CR value for the generic wildlife group (i.e. 'generic group' or 'subcategory') within which the Reference Animal fits for a given element.

(b) CR value derived from a related Reference Animal or related generic wildlife group for a given element.

(c) CR value for the given Reference Animal for an element of similar biogeochemistry.

(d) CR value for biogeochemically similar elements for encompassing or related generic wildlife group.

(e) Allometric relationships or other modelling approach.

(f) Expert judgement excluding the approaches explicitly noted above and including data derived from published reviews.

radionuclides between ambient freshwater and trout unless contact times have been protracted. Essentially, all CR data for Reference Trout have been derived either from direct empirical data for this specific RAP, or via reference to CR data for the generic group 'Fish' or CR values based upon biogeochemical analogues for trout.

4.4. Concentration ratio values for marine Reference Animals and Plants and their applicability

(106) CR data for adult marine RAPs are presented in Table 4.4. These data are based on the detailed tables, including full references, reported in Annex A.

(107) Data are available for approximately half of the elements considered in this review for Reference Brown Seaweed. The number of elements for Reference Flatfish is limited to 11, and falls to just six elements for Reference Crab. In most cases, surrogate values were derived from the generic wildlife groups. The recommended CR values compiled within IAEA (2004) were also used in a few cases, as were estuarine CR data primarily from Japanese coastal environments and the Baltic Sea.

(108) For brown seaweed, radionuclides incorporated into the thallus are absorbed directly from seawater. As seawater comprises the predominant source of elements and radionuclides to seaweed, and there appears to be little regulation of concentrations within the organism, CR values clearly constitute an appropriate measure of transfer. Very little recourse to surrogate values was required in the case of Reference Brown Seaweed, but where additional information was required, reference was made to the generic wildlife group 'Macro-algae' that includes data for red and green seaweeds.

(109) For the adult crab, most elements are acquired primarily through the ingestion of food, and equilibrium may not be attained over protracted time periods, as demonstrated by studies of technetium by marine crustaceans (Smith et al., 1998; Olsen and Vives i Batlle, 2003). Application of CRs in such cases may thus require some degree of caution. In the short term, relative to processes involving uptake and depuration, many radionuclides will adsorb on to the crustacean exoskeleton which may be an important source of radiation exposure for radionuclides emitting beta and low-energy gamma radiations, although the shell will effectively shield the living organism from lower-energy radiation emissions. The empirical database collated within the present work shows that there are few data on the assimilation of radionuclides by crab shells, the majority of data having been derived from muscle and hepatopancreas. It would be useful to collate more information on the association of radionuclides with crustacean exoskeletons, in order to further examine the importance of this exposure pathway (although this would require more complex dosimetric models than those currently being used) (ICRP, 2008). A large portion of the CR-derived values for Reference Crab have been extracted from the subcategory and generic wildlife groups 'Crustaceans – Large' and 'Crustaceans'.

Table 4.4. Concentration ratio (CR) values [geometric mean, arithmetic mean (n<2) or best-estimate-derived value in units of Bq/kg fresh weight/l water] for adult marine Reference Animals and Plants.

Element	Flatfish	Crab	Brown Seaweed
Ag	8.1E+03(a)	2.0E+05(f)	1.9E+03
Am	1.9E+02	5.0E+02(a)	7.7E+01
Ba	9.6E+00(g)	8.0E+02(g)	1.6E+03(g)
C	1.2E+04(a)	1.0E+04(a)	8.0E+03(a)
Ca	4.0E−01	4.5E+00(g)	3.8E+00(g)
Cd	1.3E+04(a)	1.2E+04	1.6E+03
Ce	2.1E+02(a)	1.0E+02(a)	9.5E+02
Cf	1.9E+02(c)	5.0E+02(d)	7.7E+01(c)
Cl	6.2E−02(g)	5.6E−02(a)	7.3E−01(a)
Cm	1.9E+02(c)	5.0E+02(d)	8.4E+03
Co	3.3E+02	4.7E+03(a)	6.8E+02
Cr	2.0E+02(f)	2.8E+02(g)	3.5E+02(g)
Cs	3.6E+01	1.4E+01	1.2E+01
Eu	7.3E+02(a)	2.4E+04(g)	1.1E+03(a)
H	1.0E+00(f)	1.0E+00(f)	3.7E−01
I	9.0E+00(f)	3.0E+00(f)	1.4E+03(a)
Ir	2.0E+01(f)	1.0E+02(f)	1.0E+03(f)
La	2.1E+02(d)	1.0E+02(d)	5.9E+03(g)
Mn	2.5E+02	2.5E+03(a)	1.1E+04
Nb	3.0E+01(f)	1.0E+02(a)	8.1E+01
Ni	2.7E+02	9.1E+02(g)	2.0E+03
Np	2.1E+01(c)	1.1E+02(a)	5.4E+01
P	9.5E+04(a)	3.0E+04(f)	9.6E+03(a)
Pa	5.0E+01(f)	1.0E+01(f)	1.0E+02(f)
Pb	3.3E+03	3.4E+03(a)	2.0E+03
Po	1.2E+04(a)	4.2E+03(a)	7.1E+02(a)
Pu	2.1E+01	3.8E+01	2.4E+03
Ra	6.3E+01(a)	7.3E+01(a)	4.4E+01(a)
Ru	1.6E+01(a)	1.0E+02(f)	2.9E+02
S	1.0E+00(f)	1.8E+00	2.4E+00(a)
Sb	6.0E+02(f)	3.0E+02(f)	1.5E+03
Se	1.0E+04(f)	1.0E+04(f)	2.0E+02(a)
Sr	1.0E+01	2.4E+00	4.3E+01
Tc	8.0E+01(f)	1.9E+02	3.7E+04
Te	1.0E+03(f)	1.0E+03(f)	1.0E+04(f)
Th	1.3E+03(a)	1.0E+03(f)	2.4E+03(a)
U	4.0E+00(a)	6.2E+00(g)	2.9E+01
Zn	2.2E+04	3.0E+05(f)	1.3E+04(g)
Zr	5.2E+01	4.9E+01(a)	6.3E+02

Shaded values were derived using surrogate CR values and according to the code given in brackets.

(a) CR value for the generic wildlife group (i.e. 'generic group' or 'subcategory') within which the Reference Animal or Plant fits for a given element.

(b) CR value derived from a related Reference Animal or Plant or related generic wildlife group for a given element.

(c) CR value for the given Reference Animal or Plant for an element of similar biogeochemistry.

(d) CR value for biogeochemically similar elements for encompassing or related generic wildlife group.

(e) Allometric relationships or other modelling approach.

(f) Expert judgement excluding approaches explicitly noted here and including data derived from published reviews.

(g) CR for the generic wildlife group (i.e. 'generic group' or 'subcategory') within which the Reference Animal or Plant fits for a given element from the estuarine environment.

(110) Uptake by adult flatfish occurs via ingestion and, for some radionuclides, via direct uptake from water over the gill surfaces. The relative importance of these factors depends on the radionuclide of interest. The CR values for Reference Flatfish collated within this report are likely to give a reasonable first indication of transfer from seawater to the organism, where it can be established that ambient water activity concentrations are not fluctuating substantially with time. However, it should be noted that for some radionuclides, such as actinides, where turnover rates in the body are slow, the CR approach has limitations. A comprehensive review of the uptake of radionuclides by marine fish was presented by Pentreath (1977a). Comprehensive data sets were available for the marine wildlife group subcategory 'Fish – Benthic Feeding' and 'Fish', and in most cases, it was possible to use these data to provide surrogate CR values for Reference Flatfish.

4.5. Transfer factor data for different life stages of development for Reference Animals and Plants

(111) Few CR data were found for the various life stages of RAPs. In view of the lack of available information, it was considered premature, if not impracticable, to attempt to derive values for each and every life stage–element combination. For this reason, CR values have not been provided in this report. However, until such values can be derived, the following recommendations have been developed for use in determining life stage CR values.

(112) Transfer to a bee colony might be considered to be similar to that for the individual adult bee, but the colony consists of different life stages, plus the non-living components of the nest within which bees live. As it forms an integral part of the colony being used as a food source for larvae and bees, transfer to and activity concentrations within the honey may provide useful information in relation to exposure estimates, notably in terms of external dose quantification.

(113) With no detailed empirical information on transfer to earthworm eggs, transfer data for the adult earthworm may provide suitable surrogate CR data, although this assumption should be tested.

(114) For duck eggs (probably the most radiosensitive stage for this Reference Animal), virtually the entire elemental/radionuclide content will have been derived from its female parent. In such cases, therefore, it may be more appropriate to relate the concentrations of radionuclides in the egg to those in the female parent. The ratio between activity concentrations in poultry meat to those in eggs for the particular radioisotope being considered could be extracted from relevant literature sources (e.g. Fesenko et al., 2009) to derive CR values for duck eggs. Such an approach was used within the derivation of values for the ERICA tool transfer database (Beresford et al., 2008a) using data for poultry from IAEA (2010). This ratio could then be applied to the CR values for adult duck. Recent work by Fesenko et al. (2009) includes a fairly comprehensive data set for poultry, but there will be many elements for which there are no data. Application of data for biogeochemically similar elements could be considered as a means of deriving a CR value for radionuclides lacking specific data.

(115) For deer calf, adult transfer data might provide reasonable proxy values if no direct empirical data are available. Results for unborn lambs have shown that Cs activity concentrations were approximately the same as those for adults. Furthermore, the resultant tissue activity concentrations of lambs and adult sheep fed herbage contaminated with ^{60}Co, ^{95}Nb, ^{106}Ru, ^{134}Cs, ^{137}Cs, ^{238}Pu, $^{239,240}Pu$, and ^{241}Am for the same time period were similar (Beresford et al., 2007). Alternatively, biokinetic models using milk (and herbage intake if the model is used to derive values for the complete period of lactation) as an intake source might be developed (although consideration of how to estimate deer milk concentrations and an initial activity concentration in the calf would be required).

(116) The frog eggs and mass of spawn are considered to be the same for the purposes of transfer. Although some data in relation to frogspawn and tadpoles exist for some radionuclides (Ophel and Fraser, 1973; Yankovich, pers. comm.), these are extremely limited. The CRs for biogeochemically similar elements might be used as surrogate values where no data exist for the radioisotope being considered. There are likely to be more data on fish egg:fish activity CRs, and these ratios might be applied to frog whole-body CR values to provide an estimate of transfer for frog eggs (e.g. Yankovich, 2009), but this would require some work to validate this as an approach.

(117) Some empirical data for frog tadpoles are available, but these highlight how trophic position may be important with differences in radionuclide CR values between life stages. For example, tadpoles are important primary consumers in aquatic ecosystems, and as adults, they become secondary consumers. It is known that ^{60}Co is assimilated by primary producers at the base of the food chain, and is quickly utilised by these organisms and depleted with increasing trophic position (Ophel and Fraser, 1973). This can lead to differences in CRs between tadpoles and adult amphibians. Ophel and Fraser (1973) have reported ^{60}Co CR values of 250 and 50 for tadpoles and bullfrogs, respectively, from Perch Lake, Ontario.

(118) Equilibration times for trout eggs and larvae are likely to be much shorter than for the adult, but the almost complete lack of data on transfer to these life stages renders any derivation of CR values inappropriate. The process of adsorption is likely to be of importance for these life stages. For trout eggs, some data are available for tissue to egg ratios for some freshwater species of teleost fish. These conversion factors could be applied to the CRs for adult trout. In cases where egg CR values are available for biogeochemically similar elements, these values might also be used as a reasonable surrogate. The adult CR value may provide a first approximation for the egg CR, although there are potential issues in applying this approach. For example, one might expect that the longer the egg remains in the water (i.e. the older it becomes), the more it will be influenced by the element/radionuclide concentrations in the water compared with the parent fish. In a specific example, Jeffree et al. (2008) found that the accumulatory and kinetic characteristics of the egg-case for some marine chondrichthian species led to enhanced exposures of embryos to certain radioisotopes. Although the trout is a teleost with a quite different egg composition and structure, the point that adult and egg may exhibit quite divergent uptake of contaminants is still pertinent and needs to be investigated further.

(119) The adsorption of radionuclides to the surface of crab eggs and larvae is an important process. For many radioisotopes, exchanges between the ambient seawater and incorporation within the organism at this stage of development will strongly influence internal activity concentrations. The larval stages of crab, known as the zoea and the megalopa, are minute organisms that swim and feed as part of the plankton. Evidence from various studies on organisms with dimensions commensurate with crab eggs and larvae suggests that equilibration occurs relatively rapidly (e.g. Stewart and Fisher, 2003; Brown et al., 2004), and thus under conditions where seawater concentrations remain constant with time, the CR approach might be expected to produce reasonable predictions of transfer. CR data for zooplankton have been published previously (e.g. IAEA, 2004; Hosseini et al., 2008), and might serve as suitable surrogate values for calculating transfer to these life stages of the crab. The female crab may carry the eggs beneath her for many months.

(120) The processes leading to the exposure of eggs and larvae of flatfish are likely to be the same as those for trout and crab, with adsorption playing an important role, but the eggs are assumed to be pelagic for flatfish and laid on a gravelly substrate for trout.

4.6. Distributions of radionuclides within the tissues of Reference Animals and Plants

(121) The Commission has noted that, for the purpose of relating dose received to the biological endpoints of interest, the critical information required for alpha particles and low-energy electrons is the concentration of the relevant radionuclide in the 'tissue or organ of interest' (ICRP, 2008). For animals, these tissues or organs of interest would appear to be the reproductive organs, as reproduction is a primary biological endpoint of interest (especially with respect to the maintenance of populations), and accumulating organs or tissues, because clearly the highest exposures will be associated with these body compartments. For plants, the tissues of relevance may be the active growing points of the shoot and root tips, the ring of phloem and xylem underneath the bark (because much of the centre of the tree trunk is dead wood), the seeds (within pine cones), and the root mass beneath the soil surface (ICRP, 2008). The Commission has started the process of considering the relative dosimetry of internal organs, such as the liver and gonads of Reference Deer, but initially for illustrative purposes rather than as definitive models (ICRP, 2008). Many marine algae also have actively growing areas and older more permanent parts, as well as reproductive structures and gametes that may be of importance in a similar way as described for plants. Thus, there is a requirement to provide information on transfer to specific tissues within RAPs for those radionuclides where alpha or beta emissions predominate or, where this is not practicable, to derive information in relation to internal distributions of these radionuclides.

(122) Whole-body CRs have been widely used in models associated with assessing the environmental impacts of radioactivity in a regulatory context (e.g. Copplestone et al., 2003; Brown et al., 2008). This partly reflects the consideration that, because a large proportion of dose–effect relationships from laboratory investigations are whole-body exposures, the most appropriate dose rates to consider are those

associated with the entire organism (Andersson et al., 2009). Nonetheless, it is recognised that for radionuclides emitting relatively short-range radiations (such as alpha particles and low-energy beta radiations) and for organisms above a certain size and complexity, doses to radiosensitive tissues are likely to dictate the resultant radiation effect. The dependence on radionuclide concentrations in that particular tissue, which can be different from the average concentration in the body, might therefore be critical. A detailed analysis of heterogeneity of radionuclides and its implications for dose in relation to a small number of examples would be useful, as has been discussed by Ulanovsky et al. (2008), and this is currently being examined by the Commission.

(123) Although it is possible to categorise data in terms of organ/body parts within the Wildlife Transfer Database, it was considered premature to report these data as organ-specific CR values. However, this information was used, along with conversion factors, to derive whole-body CRs from the organ and body-part data, and these were included in the CR values for the adult RAPs. The conversion factors used have been published (Yankovich et al., 2010) and could be used to calculate organ- or body-part-specific CR values in the future. However, more work is needed to establish a comprehensive data set of organ or body-part CR values.

(124) Studies have been conducted for the purpose of characterising the internal distribution of elements (and radioisotopes) in organisms that fall within the RAP categories. The experimental work of Pentreath (1973a,b, 1977b,c) provides information on how such experiments might be conducted, and the data that can be obtained for organ-specific transfer and internal redistributions over time. By way of example, information concerning organ-specific transfer for selected elements in Reference Flatfish is provided in Annex C. This annex also contains some information on the accumulation of radionuclides by eggs and larvae, and on biological half-lives for adult fish. These data are relevant to emergency exposure calculations.

4.7. Addressing the data gaps in the Reference Animal and Plant concentration ratio values

(125) The previous sections in this report have shown that some information is available on the transfer of some radionuclides for some RAPs, but very limited information is available on their life stages. The available information on this subject reported in the open literature has usually been described in the form of equilibrium-based CRs. It is recognised that there are a number of limitations with the application of CRs (as described above). Furthermore, there are many data gaps associated with a number of RAP–element combinations. Although approaches for filling these gaps have been proposed and used in this report to account for the lack of empirical data, the confidence associated with whether the surrogate CR values are reasonable predictors of the 'true' values for RAPs varies. In cases where the surrogate value is based on large empirical data sets for taxonomically similar organisms, and where physiological or ecological reasons to expect differences between the two groups of organisms are absent, confidence that these values represent CRs for RAPs might be considered high. Conversely, derived values based on few empirical data, for more

distantly related organisms, or approaches that are reliant on untested models, might be viewed with less confidence. On balance, the substantial reliance on derived values does not constitute a long-term solution, and alternatives should be sought for use with the RAPs.

(126) At the current time, however, the Commission believes that, given the current state of knowledge, the CR approach and the associated data-gap-filling approaches described in this report will have to provide the initial data set for the transfer of radionuclides to the RAPs. This will allow the Commission to continue to develop its RAP approach, but with recognised limitations regarding the derivation of the CRs for those cases where direct measurements of radionuclides in the environment are not available.

(127) The Commission recognises that there may be more appropriate means of obtaining transfer data for the RAPs and their life stages to provide an internally consistent (i.e. in terms of compatibility with other parts of the RAP approach) and complete data set for different tissues. Such a complete set of CR values would provide a key aspect of the radiation protection framework, and the Commission recommends that such work be undertaken. In a similar way, radiological assessment methodologies for humans have evolved over many years, and the Commission expects that the approach set out in this report will be developed and refined as additional knowledge and understanding is gained.

(128) One possible approach is to identify a series of sites where samples of each RAP, and their different life stages, could be collected and analysed. At each site, all the samples should come from the same (known and co-ordinated) location (e.g. the duck, frog, and trout should all come from the same lake). An appropriate number of samples of each RAP and their life stages should be collected, along with corresponding samples of media (water, soil). The number and specific location of any media samples would need to be taken into account, and spatial aspects, such as the home range of the RAP (and its life stages), as identified in ICRP (2008), also need to be addressed. Consideration should also be given to the timing of the sample collection. Whilst these sites could provide relevant data for the RAPs, the data will be, clearly, site specific in nature. However, these site-specific CR values could be compared with the wider CR data that are available (such as the values collated in this report for the RAPs) to help understand how CRs may vary between different geographic areas. In other words, they would serve as 'points of reference'.

(129) For each of the adult RAPs, the composition of the 40 elements should be determined for a number of the tissues of interest and associated samples of media. The tissues might include the gonads (as reproduction is a key endpoint when considering possible effects on populations of non-human species), muscle, liver, meristem, seeds, and so on, depending upon the specific RAP in question. In this way, data on elemental composition and transfer for RAPs can be collated in a more rigorously structured way than has previously been achievable through the collation of data from a multitude of often disparate and ad-hoc studies that were not designed for this purpose. During the development of the radiological protection system for humans, the Commission gathered data on the elemental composition of the human body in a similar way, and has used this information to better understand the

relationship between internal organ concentrations, the associated doses, and the biological effects. By deriving a set of compositional and transfer data for different tissues of the RAPs, it will be possible to evaluate more fully how their internal exposure is related to the radionuclide concentrations within the surrounding environment. Furthermore, alternative modelling approaches to the standard soil-, air- or water-based CRs might be explored, such as methods that account for intermediate links in the food chain by considering the diet and its elemental composition for the studied animals.

4.8. References

Andersson, P., Garnier-Laplace, J., Beresford, N.A., et al., 2009. Protection of the environment from ionising radiation in a regulatory context (PROTECT): proposed numerical benchmark values. J. Environ. Radioact. 100, 1100–1108.

Baker, A., Brooks, R., Reeves, R., 1988. Growing for gold ... and for copper ... and zinc. New Scient. 117, 44–48.

Beresford, N.A., Howard, B.J., Mayes, R.W., et al., 2007. The transfer of radionuclides from saltmarsh vegetation to sheep tissues and milk. J. Environ. Radioact. 98, 36–49.

Beresford, N.A., Barnett, C.L., Howard, B.J., et al., 2008a. Derivation of transfer parameters for use within the ERICA tool and the default concentration ratios for terrestrial biota. J. Environ. Radioact. 99, 1393–1407.

Beresford, N.A., Barnett, C.L., Jones, D.G., et al., 2008b. Background exposure rates of terrestrial wildlife in England and Wales. J. Environ. Radioact. 99, 1430–1439.

Blaylock, B.G., 1982. Radionuclide data bases available for bioaccumulation factors for freshwater biota. Nucl. Saf. 23, 428–439.

Brown, J., Børretzen, P., Dowdall, M., et al., 2004. The derivation of transfer parameters in the assessment of radiological impacts to Arctic marine biota. Arctic 57, 279–289.

Brown, J.E., Alfonso, B., Avila, R., et al., 2008. The ERICA tool. J. Environ. Radioact. 99, 1371–1383.

Copplestone, D., Wood, M.D., Bielby, S., et al., 2003. Habitat Regulations for Stage 3 Assessments: Radioactive Substances Authorisations. R&D Technical Report P3-101/SP1a. Environment Agency, Bristol.

Fesenko, S., Howard, B.J., Isamov, N., et al., 2009. Review of Russian-language studies on radionuclide behaviour in agricultural animals: part 4. Transfer to poultry. J. Environ. Radioact. 100, 815–822.

Fleishman, D.G., Nikiforov, V.A., Saulus, A.A., et al., 1994. ^{137}Cs in fish of some lakes and rivers of the Bryansk Region and north-west Russia in 1990–1992. J. Environ. Radioact. 24, 145–158.

Hosseini, A., Thørring, H., Brown, J.E., et al., 2008. Transfer of radionuclides in aquatic ecosystems – default concentration ratios for aquatic biota in the ERICA tool assessment. J Environ. Radioact. 99, 1408–1429.

Hosseini, A., Beresford, N.A., Brown, J.E., et al., 2010. Background dose-rates to reference animals and plants arising from exposure to naturally occurring radionuclides in aquatic environments. J. Radiol. Prot. 30, 235–264.

IAEA, 2001. Present and Future Environmental Impact of the Chernobyl Accident. IAEA-TECDOC-1240. International Atomic Energy Agency, Vienna.

IAEA, 2004. Sediment Distribution Coefficients and Concentration Factors for Biota in the Marine Environment. IAEA Technical Reports Series No. 422. International Atomic Energy Agency, Vienna.

IAEA, 2010. Handbook of Parameter Values for the Prediction of Radionuclide Transfer in Terrestrial and Freshwater Environments. IAEA Technical Report Series No. 472. International Atomic Energy Agency, Vienna.

IAEA, in preparation. Handbook of Parameter Values for the Prediction of Radionuclide Transfer to Wildlife. IAEA Technical Report Series. International Atomic Energy Agency, Vienna.

ICRP, 2008. Environmental protection: the concept and use of reference animals and plants. ICRP Publication 108. Ann. ICRP 38(4–6).

Jeffree, R.A., Oberhansli, F., Teyssie, J.-L., 2008. The accumulation of lead and mercury from seawater and their depuration by eggs of the spotted dogfish Scyliorhinus canicula (Chondrichthys). Arch. Environ. Contam. Toxicol. 55, 451–461.

Koulikov, A.O., Meili, M., 2003. Modelling the dynamics of fish contamination by Chernobyl radiocaesium: an analytical solution based on potassium mass balance. J. Environ. Radioact. 6, 309–326.

Kolehmainen, S., Hasanen, E., Miettinen, J.K., 1966. ^{137}Cs levels in fish of different limnological types of lakes in Finland during 1963. Health Phys. 12, 917–922.

Millero, F.J., 1996. Chemical Oceanography, 2nd ed. CRC Press, Inc.

Nagy, K.A., 2001. Food requirements of wild animals: predictive equations for free-living mammals. reptiles and birds. Nutr. Abs. Rev. Ser. B. 71, 21–31.

Nahmani, J., Hodson, M.E., Black, S., 2007. A review of studies performed to assess metal uptake by earthworms. Environ. Pollut. 145, 402–424.

Olsen, Y.S., Vives i Batlle, J., 2003. A model for the bioaccumulation of ^{99}Tc in lobsters (Homarus gammarus) from the West Cumbrian coast. J. Environ. Radioact. 67, 219–233.

Ophel, I.M., Fraser, C.D., 1973. The fate of ^{60}Co in a natural freshwater ecosystem. In: Nelson, D.J. (Ed.), Radionuclides in Ecosystems. Proceedings of the Third National Symposium on Radioecology. CONF-710501. Oak Ridge National Laboratory, Oak Ridge, TN, pp. 323–327.

Pentreath, R.J., 1973a. The accumulation of ^{65}Zn and ^{54}Mn by the plaice, Plueronetces platessa L. J. Exp. Mar. Biol. Ecol. 12, 1–18.

Pentreath, R.J., 1973b. The accumulation of ^{59}Fe and ^{58}Co by the plaice, Plueronetces platessa L. and the thornback ray, Raja clavata L. J. Exp. Mar. Biol. Ecol. 12, 315–326.

Pentreath, R.J., 1977a. Radionuclides in marine fish. Oceanogr. Mar. Biol. Ann. Rev. 15, 365–446.

Pentreath, R.J., 1977b. The accumulation of cadmium by the plaice, Plueronetces platessa L. and the thornback ray, Raja clavata L. J. Exp. Mar. Biol. Ecol. 30, 223–232.

Pentreath, R.J., 1977c. The accumulation of ^{110m}Ag by the plaice, Plueronetces platessa L. and the thornback ray, Raja clavata L. J. Exp. Mar. Biol. Ecol. 29, 315–325.

Pröhl, G., 2009. Interception of dry and wet deposited radionuclides by vegetation. J. Environ. Radioact. 100, 675–682.

Shaw, G., Bell, J.N.B., 1991. Competitive effects of potassium and ammonium on cesium uptake kinetics in wheat. J. Environ. Radioact. 13, 283–296.

Smith, D.L., Knowles, J.F., Winpenny, K., 1998. The accumulation, retention and distribution of ^{95m}Tc in crab (Cancer pagurus L.) and lobster (Homarus gammarus L.): a comparative study. J. Environ. Radioact. 40, 113–135.

Stewart, G.M., Fisher, N.S., 2003. Bioaccumulation of polonium-210 in marine copepods. Limnol. Oceanogr. 48, 2011–2019.

Ulanovsky, A., Proehl, G., Gomez-Ros, J.M., 2008. Methods for calculating dose conversion coefficients for terrestrial and aquatic biota. J. Environ. Radioact. 99, 1440–1448.

Velasco, H., Juri Ayub, J., Sansone, U., 2009. Influence of crop types and soil properties on radionuclide soil-to-plant transfer factors in tropical and subtropical environments. J. Environ. Radioact. 100, 733–738.

Watson, J.W., McAllister, K.R., Pierce, D.J., 2003. Home ranges, movements, and habitat selection of oregon spotted frogs (Rana pretiosa). J. Herpetol. 37, 292–300.

Yankovich, T.L., 2009. Mass balance approach to estimating radionuclide loads and concentrations in edible fish tissues using stable analogues. J. Environ. Radioact. 100, 795–801.

Yankovich, T.L., Beresford, N.A., Wood, M.D., et al., 2010. Whole body to tissue-specific concentration ratios for use in biota dose assessments for animals. Radiat. Environ. Biophys. 49, 549–565.

ANNEX A. DETAILED STATISTICAL INFORMATION ON CONCENTRATION RATIOS FOR REFERENCE ANIMALS AND PLANTS

A.1. Terrestrial ecosystems

Note that the geometric mean values in the following tables are approximations.

Table A.1. Wild Grass (*Poaceae*): concentration ratio values (units of Bq/kg fresh weight per Bq/kg dry weight).

Element	Arithmetic mean	Arithmetic standard deviation	Geometric mean	Geometric standard deviation	n	Ref ID
Am	2.8E−01	4.4E−01	1.5E−01	3.1E+00	23	486
Cd	2.8E+00	2.9E−01	2.7E+00	1.1E+00	200	202
Cl	5.4E+01	2.4E+01	4.9E+01	1.5E+00	8	494
Cs	1.8E+00	3.2E+00	8.6E−01	3.3E+00	1068	210, 253, 272, 395, 409, 413, 414, 448, 453, 486, 501, 510, 519
Ni	2.2E−01	1.6E−01	1.8E−01	1.9E+00	58	285, 286, 334
Pb	1.3E−01	1.9E−01	7.5E−02	2.9E+00	72	220, 293, 334
Po	4.2E−01	6.3E−01	2.3E−01	3.0E+00	22	220, 334
Pu	4.3E−02	3.6E−02	3.3E−02	2.1E+00	5	486
Ra	3.2E−01	1.1E+00	9.2E−02	4.8E+00	168	220, 266, 272, 273, 287, 288, 292, 293, 334, 459
Sb	4.1E+01				1	194
Se	1.8E+00	1.6E+00	1.3E+00	2.1E+00	48	497
Sr	2.4E+00	2.1E+00	1.7E+00	2.2E+00	36	163, 414, 451, 486, 501
Tc	4.7E+00	5.0E+00	3.2E+00	2.4E+00	6	486
Th	1.5E−01	1.8E−01	9.5E−02	2.6E+00	30	272, 334, 430, 459
U	1.6E−01	5.4E−01	4.3E−02	5.0E+00	151	220, 266, 272, 279, 292, 334, 430, 457, 459, 489
Zn	3.5E+00	3.2E+00	2.6E+00	2.2E+00	6	334

Table A.2. Pine Tree (*Pinaceae*): concentration ratio values (units of Bq/kg fresh weight per Bq/kg dry weight).

Element	Arithmetic mean	Arithmetic standard deviation	Geometric mean	Geometric standard deviation	*n*	Ref ID
Ba	1.9E–01	1.3E–01	1.6E–01	1.8E+00	3	467
Ce	3.3E–03				2	467
Cl	1.5E+00	1.4E+00	1.1E+00	2.2E+00	5	251
Co	1.7E–03	1.1E–03	1.4E–03	1.9E+00	3	467
Cr	4.1E–03	1.8E–03	3.8E–03	1.5E+00	3	467
Cs	1.5E–01	2.5E–01	7.5E–02	3.2E+00	235	183, 472, 474, 475, 476, 484
Eu	2.1E–03				2	467
La	3.5E–03	1.8E–03	3.1E–03	1.6E+00	3	467
Pb	6.1E–02	3.4E–02	5.3E–02	1.7E+00	10	220
Po	4.7E–02	2.8E–02	4.0E–02	1.7E+00	10	220
Ra	9.2E–04	9.9E–04	6.3E–04	2.4E+00	10	220
Sr	5.6E–01	1.4E+00	2.0E–01	4.1E+00	77	467, 479, 480, 482, 484
Th	7.2E–04	1.5E–03	3.2E–04	3.6E+00	5	200, 467
U	1.3E–03	1.0E–03	9.9E–04	2.0E+00	13	200, 220
Zn	3.8E–02	1.7E–02	3.5E–02	1.5E+00	3	467

Table A.3. Earthworm (*Lumbricidae*): concentration ratio values (units of Bq/kg fresh weight per Bq/kg dry weight).

Element	Arithmetic mean	Arithmetic standard deviation	Geometric mean	Geometric standard deviation	*n*	Ref ID
Am	1.1E+00				1	486
Cd	4.6E+00	3.6E+00	3.6E+00	2.0E+00	398	199, 229, 264, 344
Ce	3.7E–04				1	264
Cl	1.8E–01	6.0E–02	1.7E–01	1.4E+00	17	238
Cs	5.0E–02	1.5E–02	4.8E–02	1.3E+00	7	207, 264
Eu	7.9E–04				1	264
I	1.6E–01	6.7E–02	1.4E–01	1.5E+00	10	238
Mn	1.6E–02	9.1E–03	1.3E–02	1.7E+00	5	199, 264
Nb	5.1E–04				1	264
Ni	2.4E–02	6.1E–03	2.3E–02	1.3E+00	5	199, 264
Pb	8.0E–01	8.1E–01	5.7E–01	2.3E+00	409	159, 199, 229,264, 344
Po	1.0E–01	3.9E–02	9.6E–02	1.4E+00	7	384
Sb	6.0E–03				1	264
Se	1.5E+00				1	231
Sr	9.0E–03				1	264
U	8.8E–03				1	264
Zn	4.0E+00	1.6E+00	3.7E+00	1.5E+00	383	344

Table A.4. Bee (*Apidea*): concentration ratio values (units of Bq/kg fresh weight per Bq/kg dry weight).

No empirical data.

Table A.5. Frog (*Ranidae*): concentration ratio values (units of Bq/kg fresh weight per Bq/kg dry weight).

Element	Arithmetic mean	Arithmetic standard deviation	Geometric mean	Geometric standard deviation	*n*	Ref ID
Am	1.0E−01	2.6E−02	1.0E−01	1.3E+00	7	486
Cd	1.5E−02	7.9E−03	1.3E−02	1.7E+00	5	213
Cs	5.5E−01	9.0E−01	2.8E−01	3.2E+00	105	188, 205, 256, 486
Pb	3.1E−03	2.2E−03	2.6E−03	1.9E+00	6	213
Sr	1.5E+00	1.4E+00	1.1E+00	2.2E+00	14	188, 486

Table A.6. Duck (*Anatidae*): concentration ratio values (units of Bq/kg fresh weight per Bq/kg dry weight).

Element	Arithmetic mean	Arithmetic standard deviation	Geometric mean	Geometric standard deviation	*n*	Ref ID
Am	3.2E−02	1.6E−02	2.8E−02	1.6E+00	3	486
Cs	4.5E−01	7.8E−01	2.2E−01	3.2E+00	40	163, 190, 263, 486
Pu	1.1E−02	4.8E−03	1.0E−02	1.5E+00	5	486
Ra	8.4E−02	9.7E−02	5.5E−02	2.5E+00	5	239
Sr	1.3E−01	1.0E−01	1.1E−01	2.0E+00	4	190, 263, 486
Tc	1.7E−01				2	486

Table A.7. Rat (*Muridae*): concentration ratio values (units of Bq/kg fresh weight per Bq/kg dry weight).

Element	Arithmetic mean	Arithmetic standard deviation	Geometric mean	Geometric standard deviation	*n*	Ref ID
Am	3.7E−04	9.9E−05	3.6E−04	1.3E+00	9	488
Co	3.0E−01	3.7E−01	1.8E−01	2.7E+00	29	161
Cs	9.9E−01	4.3E+00	2.2E−01	5.6E+00	70	268, 405, 486, 488
Pb	1.5E−02	1.7E−02	9.6E−03	2.5E+00	36	211
Po	7.5E−04				1	450
Pu	7.9E−02	3.2E−01	1.9E−02	5.4E+00	27	268, 405, 488
Ra	4.7E−02	1.8E−02	4.4E−02	1.4E+00	5	260
Sr	3.0E+00	2.9E+00	2.2E+00	2.2E+00	37	268, 405
Th	6.3E−05				1	450
U	6.5E−04				1	450

Table A.8. Deer (*Cervidae*): concentration ratio values (units of Bq/kg fresh weight per Bq/kg dry weight).

Element	Arithmetic mean	Arithmetic standard deviation	Geometric mean	Geometric standard deviation	*n*	Ref ID
Am	7.5E−03	2.6E−02	2.1E−03	4.9E+00	13	184
Cs	4.1E+00	9.4E+00	1.6E+00	3.9E+00	1745	163, 184, 190, 208, 209, 228, 230, 294
Pu	2.6E−03	7.2E−03	8.9E−04	4.3E+00	15	184, 222
Sr	2.9E+00	2.8E+00	2.1E+00	2.3E+00	58	163, 190, 228

Table A.9. References for terrestrial Reference Animals and Plants (Tables A.1–A.8).

Ref ID	Reference	Ref ID	Reference
159	Andrews et al. (1989)	287	Mislevy et al. (1989)
161	Bastian and Jackson (1975)	288	Mortvedt (1994)
163	Beresford et al. (2005)	292	Rumble and Bjugstad (1986)
183	Ertel and Ziegler (1991)	293	Simon and Fraley (1986)
184	Ferenbaugh et al. (2002)	294	Steinnes et al. (2009)
188	Gaschak (pers. comm.)	334	COGEMA (2000)
190	Gaschak et al. (2003)	344	Vermeulen et al. (2009)
194	Ghuman et al. (1993)	384	Brown et al. (2009)
199	Hendriks et al. (1995)	395	Bekyasheva et al. (1990)
200	Hinton et al. (2005)	405	Gashchak and Beresford (2009)
202	Hunter and Johnson (1984)	409	Grebenshchikova et al. (1992)
205	Jagoe et al. (2002)	413	Ilyin et al. (1991)
207	Janssen et al. (1996)	414	Khomich (1990)
208	Johanson (1994)	430	Miroshichenko et al. (1990)
209	Johanson and Bergstrom (1994)	448	Prister et al. (1988)
210	Johanson et al. (1994)	450	Read and Pickering (1999)
211	Johnson and Roberts (1978)	451	Sanzharova et al. (1990)
213	Karasov et al. (2005)	453	Shutov et al. (1993)
220	Mahon and Mathews (1983)	457	Titaeva (1992)
222	Mietelski (2001)	459	Titaeva and Toskaev (1983)
228	Miretsky et al. (1993)	467	Higley (2010)
229	Morgan and Morgan (1990)	472	Dvornik and Ipatyev (2005)
230	Nelin (1995)	474	Ipatyev et al. (2004a)
231	Nielsen and Gissel-Nielsen (1975)	475	Ipatyev et al. (2004b)
238	Pokarzhevskii and Zhulidov (1995)	476	Bulko and Ipatyev (2005)
239	Pokarzhevskii and Krivolutzkii (1997)	479	Mukhamedshin et al. (2000)
251	Sheppard et al. (1999)	480	Perevolotsky (2006a)
253	Tsvetnova and Sheglov (2009)	482	Perevolotsky (2006b)
256	Stark et al. (2004)	484	Shcheglov (1997)
260	Verhovskaya (1972)	486	Wood (2010)
263	Wood et al. (2008)	488	Wood et al. (2009)
264	Yoshida et al. (2005)	489	Vandenhove et al. (2006)
266	Apps et al. (1988)	494	Kashparov et al. (2007)
268	Beresford et al. (2008a)	497	Sharmasarkar and Vance (2002)
272	Dowdall et al. (2005)	501	Vidal et al. (2001)
273	Gerzabek et al. (1998)	510	Ponikarova et al. (1990)
279	Idiz et al. (1986)	519	Livens et al. (1991)
285	Mascanzoni (1989a)		
286	Mascanzoni (1989b)		

A.2. Freshwater ecosystems

Note that the geometric mean values in the following tables are approximations.

Table A.10. Trout (*Salmonidae*): concentration ratio values (units of Bq/kg fresh weight per Bq/l).

Element	Arithmetic mean	Arithmetic standard deviation	Geometric mean	Geometric standard deviation	*n*	Ref ID
Ba	1.5E+01	2.4E+01	8.1E+00	3.1E+00	87	333, 336, 376
C	1.7E+05	6.3E+05	4.3E+04	5.2E+00	39	330
Ca	5.2E+02	5.7E+02	3.5E+02	2.4E+00	124	333, 339, 343, 361, 371
Ce	4.5E+02	6.0E+02	2.7E+02	2.8E+00	66	333
Co	1.0E+02	4.5E+01	9.5E+01	1.5E+00	56	333
Cr	1.9E+02	1.4E+02	1.5E+02	1.9E+00	66	333, 343
Cs	3.6E+03	3.1E+03	2.7E+03	2.1E+00	118	146, 313, 326, 327, 332, 333, 402, 416
Eu	3.3E+01	1.6E+01	3.0E+01	1.6E+00	18	333
I	7.6E+01	5.3E+01	6.2E+01	1.9E+00	17	329, 333
La	3.0E+02	3.4E+02	2.0E+02	2.5E+00	60	333
Mn	5.0E+03	6.7E+03	3.0E+03	2.7E+00	126	333, 336, 339, 343, 361, 376
Ni	1.7E+01	1.4E+01	1.3E+01	2.0E+00	15	333, 343
P	7.4E+05	2.6E+05	7.0E+05	1.4E+00	92	333
Pb	3.9E+02	1.6E+03	9.1E+01	5.5E+00	22	336, 361, 383
Po	2.0E+02	1.6E+02	1.6E+02	2.0E+00	10	336, 343
Pu	2.6E+01	2.3E+01	2.0E+01	2.1E+00	5	306, 321
Ra	5.9E+01	8.5E+01	3.4E+01	2.9E+00	46	305, 339, 343, 361, 371
Sb	3.4E+01	1.1E+02	9.8E+00	4.8E+00	105	333, 399
Se	6.6E+03	3.2E+03	5.9E+03	1.6E+00	15	361, 371, 376
Sr	1.8E+02	2.5E+02	1.0E+02	2.9E+00	129	333, 336, 339, 361, 371, 376, 389, 416
U	2.1E+01	4.7E+01	8.5E+00	3.8E+00	36	339, 361, 371
Zn	1.1E+04	4.8E+03	1.0E+04	1.5E+00	100	333, 336, 339
Zr	5.2E+02	1.9E+02	4.9E+02	1.4E+00	4	333

Table A.11. Frog (*Ranidae*): concentration ratio values (units of Bq/kg fresh weight per Bq/l).

Element	Arithmetic mean	Arithmetic standard deviation	Geometric mean	Geometric standard deviation	*n*	Ref ID
Ca	1.2E+03	1.3E+03	8.5E+02	2.4E+00	8	333
Cr	6.5E+01				2	333
Pb	5.3E+00				2	333
Zn	7.3E+02				2	333

Table A.12. Duck (*Anatidae*): concentration ratio values (units of Bq/kg fresh weight per Bq/l).

No empirical data.

Table A.13. References for freshwater Reference Animals and Plants (Tables A.10–A.12).

Ref ID	Reference	Ref ID	Reference
146	Vakulovsky (2008)	336	Areva (2010)
305	Clulow et al. (1998)	339	COGEMA (2005)
306	Edgington et al. (1976)	343	COGEMA (1998)
313	Hewett and Jefferies (1978)	361	Cameco (2001)
321	Marshall et al. (1975)	371	Cameco (2000)
326	Preston and Dutton (1967)	376	Cameco (2005)
327	Rowan and Rasmussen (1994)	383	Saxen and Outola (2009)
329	Shorti et al. (1969)	389	Outola et al. (2009)
330	Stephenson et al. (1994)	399	Culioli et al. (2009)
332	Vanderploeg et al. (1975)	402	Dushauskene-Duzh (1969)
333	Yankovich (2010)	416	Kulikov and Chebotina (1988)

A.3. Marine ecosystems

Note that the geometric mean values in the following tables are approximations.

Table A.14. Brown Seaweed (*Fucaceae*): concentration ratio values (units of Bq/kg fresh weight per Bq/l).

Element	Arithmetic mean	Arithmetic standard deviation	Geometric mean	Geometric standard deviation	*n*	Ref ID
Ag	3.8E+03	6.3E+03	1.9E+03	3.2E+00	10	7, 16, 21, 149
Am	9.8E+01	7.7E+01	7.7E+01	2.0E+00	33	16, 381
Cd	2.0E+03	1.5E+03	1.6E+03	2.0E+00	6	97
Ce	9.7E+02	2.1E+02	9.5E+02	1.2E+00	3	114
Cm	1.1E+04	8.0E+03	8.4E+03	2.0E+00	13	35
Co	1.2E+03	1.6E+03	6.8E+02	2.8E+00	62	26, 108, 120, 149, 381
Cs	7.1E+01	4.1E+02	1.2E+01	6.5E+00	412	43, 63, 70*, 78, 90, 91, 107, 108, 109, 110, 111, 114, 120, 125, 146, 381
H	3.7E−01				13	381
Mn	1.2E+04	7.0E+03	1.1E+04	1.7E+00	10	10, 47, 120
Nb	1.3E+02	1.5E+02	8.1E+01	2.6E+00	3	120
Ni	2.0E+03	1.1E+03			2	47
Np	5.7E+01	1.9E+01	5.4E+01	1.4E+00	47	35, 86, 515
Pb	2.5E+03	1.8E+03	2.0E+03	1.9E+00	5	97
Pu	3.2E+03	2.6E+03	2.4E+03	2.1E+00	146	50, 51, 63, 68, 107, 108, 111, 127, 146, 381
Ru	3.5E+02	2.3E+02	2.9E+02	1.8E+00	3	114
Sb	1.5E+03	2.1E+03			2	89, 149
Sr	5.4E+01	4.0E+01	4.3E+01	1.9E+00	40	107, 108, 111, 118, 120, 146, 381
Tc	5.6E+04	6.2E+04	3.7E+04	2.5E+00	166	12, 23, 38, 66, 78, 89, 109, 110, 112, 381
U	2.9E+01				17	381
Zr	6.4E+02	1.2E+02	6.3E+02	1.2E+00	3	114

*Estuarine data (Baltic Sea).

Table A.15. Crab (*Cancridae*): concentration ratio values (units of Bq/kg fresh weight per Bq/l).

Element	Arithmetic mean	Arithmetic standard deviation	Geometric mean	Geometric standard deviation	*n*	Ref ID
Cd	1.2E+04	7.2E+02	1.2E+04	1.1E+00	4	514
Cs	1.7E+01	1.2E+01	1.4E+01	1.9E+00	66	78
Pu	3.8E+01				1	51
S	2.0E+00	7.2E−01	1.8E+00	1.4E+00	4	514
Sr	2.5E+00	5.4E−01	2.4E+00	1.2E+00	4	514
Tc	2.1E+02	1.1E+02	1.9E+02	1.6E+00	17	25, 78

Table A.16. Flatfish (*Pleuronectidae*): concentration ratio values (units of Bq/kg fresh weight per Bq/l).

Element	Arithmetic mean	Arithmetic standard deviation	Geometric mean	Geometric standard deviation	*n*	Ref ID
Am	3.2E+02	4.2E+02	1.9E+02	2.7E+00	23	55, 78, 116
Ca	4.0E−01				1	333
Co	4.2E+02	3.3E+02	3.3E+02	2.0E+00	6	67, 72, 147
Cs	5.6E+01	6.9E+01	3.6E+01	2.6E+00	315	62, 67, 78, 90, 99, 110, 111, 117, 125, 132, 137, 143, 145, 147, 386
Mn	2.6E+02	8.0E+01	2.5E+02	1.4E+00	6	147, 333
Ni	2.8E+02	5.3E+01	2.7E+02	1.2E+00	5	333
Pb	4.4E+03	3.7E+03	3.3E+03	2.1E+00	5	333
Pu	5.1E+01	1.1E+02	2.1E+01	3.8E+00	25	51, 55, 78, 120, 126, 145, 386
Sr	1.4E+01	1.1E+01	1.0E+01	2.1E+00	12	91, 110, 145
Zn	2.2E+04	3.0E+03	2.2E+04	1.1E+00	5	333
Zr	5.2E+01				1	83

Table A.17. References for marine Reference Animal and Plants (Tables A.14–A.16).

Ref ID	Reference	Ref ID	Reference
7	Amiard (1978)	97	Melhuus et al. (1978)
10	Ancellin et al. (1979)	99	Naustvoll et al. (1997)
12	ARCTICMAR (2000)	107	NRPA (1994)
16	Boisson et al. (1997)	108	NRPA (1997)
21	Bowen (1979)	109	NRPA (1999)
23	Brown et al. (1999)	110	NRPA (2000)
25	Busby et al. (1997)	111	NRPA (1995)
26	Buyanov and Boiko (1972)	112	NRPA (1998)
35	Coughtrey et al. (1984)	114	Pentreath (1976)
38	Dahlgaard et al. (1997)	116	Pentreath and Lovett (1978)
43	Fisher et al. (1999)	117	Pertsov (1978)
47	Foster (1976)	118	Polikarpov (1964)
50	Germain et al. (2000)	120	Polikarpov (1966)
51	Gomez et al. (1991)	125	Rissanen et al. (1997)
55	Hayashi et al. (1990)	126	Rissanen et al. (2000)
62	Holm et al. (1994)	127	Rissanen et al. (1995)
63	Holm et al. (1983)	132	Shutov et al. (1999)
66	Hurtgen et al. (1988)	137	Steele (1990)
67	Ichikawa and Ohno (1974)	143	Tateda and Koyanagi (1996)
68	Ikaheimonen et al. (1995)	145	Templeton (1959)
70	Ilus et al. (2005)	146	Vakulovsky (2008)
72	Ishii et al. (1976)	147	Van As et al. (1975)
78	Kershaw et al. (2005)	149	Van Weers and Van Raaphorst (1979)
83	Kurabayashi et al. (1980)	333	Yankovich (2010)
86	Lindahl et al. (2005)	381	Westlakes Scientific Consultants (pers. comm.)
89	Masson et al. (1995)	386	Lee (2006)
90	Matishov et al. (1999)	514	Barrento et al. (2009)
91	Matishov et al. (1994)	515	Pentreath (1981)

References

Amiard, J.C., 1978. Ag-110m contamination mechanisms in a marine benthic food chain. 3. Influence of the mode of contamination upon the distribution of the radionuclide. Helgol. Wissenschaftliche Meeresunters. 31, 444–456.

Ancellin, J., Gueguenіat, P., Germaine, P., 1979. Radioecologie Marine. Eyrolles, Paris.

Andrews, S.M., Johnson, M.S., Cooke, J.A., 1989. Distribution of trace-element pollutants in a contaminated grassland ecosystem established on metalliferous fluorspar tailings 1. Lead. Environ. Pollut. 58, 73–85.

Apps, M.J., Duke, M.J.M., Stephens-Newsham, L.G., 1988. A study of radionuclides in vegetation on abandoned uranium tailings. J. Radioanalyt. Nucl. Chem. Art. 123, 133–147.

ARCTICMAR, 2000. Radiological assessment of consequences from radioactive contamination of arctic marine areas. In: Iosjpe, M. (Ed.), Annual Progress Report 01.09.99–31.08.00. Norwegian Radiation Protection Authority, Østerås.

Areva, 2010. Shea Creek Project Area, Environmental Baseline Investigation 2007–2009. Draft Report. Canada North Environmental Services.

Barrento, S., Marques, A., Teixeira, B., et al., 2009. Accumulation of elements (S, As, Br, Sr, Cd, Hg, Pb) in two populations of Cancer pagurus: ecological implications to human consumption. Food Chem. Toxicol. 47, 150–156.

Bastian, R.K., Jackson, W.B., 1975. Cs-137 and Co-60 in a terrestrial community at Enewatak Atoll. In: Cushing, C.E.J. (Ed.), Proceedings of a Symposium on Radioecology and Energy Resources. Academic Press, London, pp. 313–320.

Bekyasheva, T.A., Shutov, V.N., Basalayeva, L.N., 1990. Effects of soil properties on radiocesium accumulation by natural grasses. Radioecology of Soil and Plants. In: 3rd All-Union Conference on Agricultural Radiology, 2–7 July 1990, Russian Institute of Agricultural Radiology and Agroecology, Obninsk, pp. 49–50.

Beresford, N.A., Wright, S.M., Barnett, C.L., et al., 2005. Approaches to estimating the transfer of radionuclides to Arctic biota. Radioprotection 40, S285–S290.

Beresford, N.A., Gaschak, S., Barnett, C.L., et al., 2008. Estimating the exposure of small mammals at three sites within the Chernobyl exclusion zone – a test application of the ERICA tool. J. Environ. Radioact. 99, 1496–1502.

Boisson, F., Hutchins, D.A., Fowler, S.W., et al., 1997. Influence of temperature on the accumulation and retention of 11 radionuclides by the marine alga Fucus vesiculosus L. Mar. Pollut. Bull. 35, 313–321.

Bowen, H.J.M., 1979. Environmental Chemistry of the Elements. Academic Press, London.

Brown, J., Gjelsvik, R., Holm, E., et al., 2009. Filling knowledge gaps in radiation protection methodologies for non-human biota. Final Summary Report. Nordic Nuclear Safety Research (NKS) Report, ISBN 978-87-7893-254-9, NKS, DK - 4000 Roskilde, Denmark, pp. 17.

Brown, J.E., Kolstad, A.K., Brungot, A.L., et al., 1999. Levels of Tc-99 in seawater and biota samples from Norwegian coastal waters and adjacent seas. Mar. Pollut. Bull. 38, 560–571.

Bulko, N.I., Ipatyev, V.A., 2005. Forest ecosystems after the accident at the Chernobyl NPP: condition, prediction, response of the population, ways of rehabilitation. In: Ipatyev, V.A. (Ed.), Forest. Human. Chernobyl. Institute of Forest of the NAS of Belarus, Gomel, pp. 225–277 (in Russian).

Busby, R., McCartney, M., Mcdonald, P., 1997. Technetium-99 concentration factors in Cumbrian seafood. Radioprot. – Colloques 32, 311–316.

Buyanov, N.I., Boiko, E.V., 1972. Cobalt-60 accumulation by brown algae of different age. Okeanologiya 12, 471–474 (in Russian).

Cameco, 2000. Current Period Environmental Monitoring Program for the Beaverlodge Mine Site – Revision 2. Conor Pacific Environmental Technologies Inc., Saskatoon, Saskatchewan.

Cameco, 2001. Technical Memorandum – Water Quality Results from Eagle Drill Hole and Dubyna Drill Hole. Canada North Environmental Services, Saskatoon, Saskatchewan.

Cameco, 2005. McArthur River Operation Comprehensive Environmental Effects Monitoring Program, Interpretative Report. Golder Associates, Saskatoon, Saskatchewan.

Clulow, F.V., Dave, N.K., Lim, T.P., et al., 1998. Radium-226 in water, sediments, and fish from lakes near the city of Elliot Lake, Ontario, Canada. Environ. Pollut. 99, 13–28.

COGEMA, 1998. Cluff Lake Project, Suspension of Operations and Eventual Decommissioning of the Tailings Management Area (TMA), Biological Environment. Conor Pacific Environmental Technologies Inc., Saskatoon, Saskatchewan.

COGEMA, 2000. Conor Pacific Environmental Technologies Inc. Cluff Lake Decommissioning Comprehensive Study for Decommissioning, Comprehensive Study Report. Sections For: Existing Environment and Assessment of Potential Impacts, COGEMA Resources Inc. Saskatoon, Saskatchewan.

COGEMA, 2005. Cluff Lake Uranium Mine, 2004. Environmental Effects Monitoring and Environmental Monitoring Programs. Canada North Environmental Services. Saskatoon, Saskatchewan.

Coughtrey, P.J., Jackson, D., Jones, C.H., et al., 1984. Radionuclide Distribution and Transport in Terrestrial and Aquatic Ecosystems – a Critical Review of Data, vol. 4. A.A. Balkema, Rotterdam/Boston.

Culioli, J-L., Fouquoire, A., Calendini, S., et al., 2009. Trophic transfer of arsenic and antimony in a freshwater ecosystem: a field study. Aquat. Toxicol. 94, 286–293.

Dahlgaard, H., Bergan, T.D.S., Christensen, G.C., 1997. Technetium-99 and caesium-137 time series at the Norwegian coast monitored by the brown alga Fucus vesiculosus. Radioprotection – Colloques 32, 353–358.

Dowdall, M., Gwynn, J.P., Moran, C., et al., 2005. Uptake of radionuclides by vegetation at a High Arctic location. Environ. Pollut. 133, 327–332.

Dushauskene-Duzh, N.R.F., 1969. A Comparative Study into Accumulation of Strontium-90 and Lead-210 in Fresh Water Hydrobionts of the Lithuanian Republic. Ph.D. Thesis. INUM, Sebastopol (in Russian).

Dvornik, A.V., Ipatyev, V.A., 2005. Modelling and predictive estimates of the radionuclide accumulation by woody plants and forest-derived foodstuffs. In: Ipatyev, V.A. (Ed.), Forest Human. Chernobyl. Forest Ecosystems after the Accident at the Chernobyl NPP: Condition, Prediction, Response of the Population, ways of Rehabilitation. Institute of Forest of National Academy of Science, Gomel, pp. 178–213 (in Russian).

Edgington, D.N., Wahlgren, M.A., Marshall, J.S., 1976. The behaviour of plutonium in aquatic ecosystems: a summary of studies on the Great Lakes. In: Miller, M.W., Stannard, J.N. (Eds.), Environmental Toxicity of Aquatic Radionuclides: Models and Mechanisms. Ann Arbor Science Publishers, Ann Arbor, Michigan, pp. 45–79.

Ertel, J., Ziegler, H., 1991. Cs-134/137 contamination and root uptake of different forest trees before and after the Chernobyl accident. Radiat. Environ. Biophys. 30, 147–157.

Ferenbaugh, J.K., Fresquez, P.R., Ebinger, M.H., et al., 2002. Radionuclides in soil and water near a low level disposal site and potential ecological and human health impacts. Environ. Monitor. Assess. 74, 243–254.

Fisher, N.S., Fowler, S.W., Boisson, F., et al., 1999. Radionuclide bioconcentration factors and sediment partition coefficients in Arctic Seas subject to contamination from dumped nuclear wastes. Environ. Sci. Technol. 33, 1979–1982.

Foster, P., 1976. Concentrations and concentration factors of heavy metals in brown algae. Environ. Pollut. 10, 45–53.

Gaschak, S., Chizhevsky, I., Arkhipov, A., et al., 2003. The transfer of Cs-137 and Sr-90 to wild animals within the Chernobyl exclusion zone. In: International Conference on the Protection of the Environment from the Effects of Ionizing Radiation, 6–10 October 2003, Stockholm. International Atomic Energy Agency, Vienna, pp. 200–202.

Gashchak, S., Beresford, N.A., 2009. Data forwarded through N. Beresford (personal communication).

Germain, P., Leclerc, G., Le Cavelier, S., et al., 2000. Évolution spatio-temporelle des concentrations, des rapports isotopiques et des facteurs de concentration du plutonium dans une espèce d'algue et deux espèces de mollusques en Manche. Radioprotection 35, 175–200.

Gerzabek, M.H., Strebl, F., Temmel, B., 1998. Plant uptake of radionuclides in lysimeter experiments. Environ. Pollut. 99, 93–103.

Ghuman, G.S., Motes, B.G., Fernandez, S.J., et al., 1993. Distribution of antimony-125, cesium-137 and iodine-129 in the soil-plant system around a nuclear fuel reprocessing plant. J. Environ. Radioact. 21, 161–176.

Gomez, L.S., Marietta, M.G., Jackson, D.W., 1991. Compilation of Selected Marine Radioecological Data for the Formerly Utilized Sites Remedial Action Program: Summaries of Available Radioecological Concentration Factors and Biological Half-lives. Sandia National Laboratories Report SAND89-1585 RS-8232-2. Sandia National Labs., Albuquerque, NM (USA).

Grebenshchikova, N.V., Firsakova, S.K., Novik, A.A., et al., 1992. Investigations of radiocaesium behaviour in soil-vegetation cover of Belorussia Polesje after the accident at the Chernobyl NPP. Agrokhimiya 1, 91–99 (in Russian).

Hayashi, N., Katagiri, H., Narita, O., et al., 1990. Concentration factors of plutonium and americium for marine products. J. Radioanal. Nucl. Chem. 138, 331–336.

Hendriks, A.J., Ma, W.C., Brouns, J.J., et al., 1995. Modelling and monitoring organochlorine and heavy metal accumulation in soils, earthworms, and shrews in Rhine-delta floodplains. Arch. Environ. Contamin. Toxicol. 29, 115–127.

Hewett, C.J., Jefferies, D.F., 1978. The accumulation of radioactive caesium from food by the plaice (Pleuronectes platessa) and the brown trout (Salmo trutta). J. Fish Biol. 13, 143–153.

Higley, K.A., 2010. Estimating transfer parameters in the absence of data. Radiat. Environ. Biophys. 49, 645–656.

Hinton, T.G., Knox, A.S., Kaplan, D.I., et al., 2005. Phytoextraction of uranium and thorium by native trees in a contaminated wetland. J. Radioanal. Nucl. Chem. 264, 417–422.

Holm, E., Persson, B.R.R., Hallstadius, L., et al., 1983. Radiocesium and transuranium elements in the Greenland and Barents Seas. Oceanol. Act. 6, 457–462.

Holm, E., Ballestra, S., Lopez, J.J., et al., 1994. Radionuclides in macro algae at Monaco following the Chernobyl accident. J. Radioanal. Nucl. Chem. 177, 51–72.

Hunter, B.A., Johnson, M.S., 1984. Food chain relationship of copper and cadmium in herbivorous and insectivorous small mammals. In: Osborn, D. (Ed.), Metals in Animals. CEH Monkswood, Huntingdon, pp. 5–10.

Hurtgen, C., Koch, G., Van Der Ben, D., et al., 1988. The determination of technetium-99 in the brown marine alga Fucus spiralis collected along the Belgian Coast. Sci. Total Environ. 70, 131–149.

Ichikawa, R., Ohno, S., 1974. Levels of cobalt, caesium and zinc in some marine organisms in Japan. Bull. Jap. Soc. Sci. Fish. 40, 501–508.

Idiz, E.F., Carlisle, D., Kaplan, I.R., 1986. Interaction between organic matter and trace metals in a uranium rich bog, Kern County, California. U.S.A. Appl. Geochem. 1, 573–590.

Ikaheimonen, T.K., Rissanen, K., Matishov, D.G., et al., 1995. Plutonium in fish, algae, and sediments in the Barents, Petshora, and Kara Sea. In: International Conference on Environmental Radioactivity in the Arctic, 21–25 August, Oslo. Norwegian Radiation Protection Authority, Østerås, pp. 227–232.

Ilus, E., Klemola, S., Ikaheimonen, T.K., et al., 2005. Indicator value of certain aquatic organisms for radioactive substances in the sea areas off the Loviisa and Olkiluoto nuclear power plants (Finland). In: Proceedings of the Summary Seminar within the NKS-B Programme 2002–2005, 24–25 October 2005, Tartu, Nordic Nuclear Safety Research (NKS), DK – 4000 Roskilde, Denmark, pp. 67–73.

Ilyin, M.I., Perepelyatnikov, G.P., Prister, B.S., 1991. Effects of radical improvement of natural meadows in the Ukrainian Poles'ye on radiocaesium transfer from soil to grass stand. Agrochimiya 1, 101–105.

Ipatyev, M.A., Bulko, N.I., Mitin, N.V., et al., 2004a. The hydromeliorative method for decreasing radionuclide concentrations in forest ecosystems. Radioecol. Phenom. Forest Ecosyst. 29, 137–166.

Ipatyev, M.A., Bulko, N.I., Mitin, N.V., et al., 2004b. The silvicultural (phytologic) method for decreasing radioactivity levels in forest ecosystems. Radioecol. Phenom. Forest Ecosyst. 37, 67–104.

Ishii, T., Suzuki, H., Iimura, M., et al., 1976. Concentration of Trace Elements in Marine Organisms. NIRS-R-5, National Institute of Radiological Sciences, Chiba, Japan, pp. 28–29.

Jagoe, C.H., Majeske, A.J., Oleksyk, T.K., et al., 2002. Radiocesium concentrations and DNA strand breakage in two species of amphibians from the Chernobyl exclusion zone. Radioprotection – Colloques 37, 873–878.

Janssen, M.P.M., Glastra, P., Lembrechts, J.F.M.M., 1996. Uptake of Cs-134 from a sandy soil by two earthworm species: the effects of temperature. Arch. Environ. Contam. Toxicol. 31, 184–191.

Johanson, K.J., 1994. Radiocaesium in game animals in Nordic countries. In: Nordic Radioecology – the Transfer of Radionuclides through Nordic Ecosystems to Man. Elsevier, Amsterdam, pp. 287–301.

Johanson, K.J., Bergstrom, R., 1994. Radiocesium transfer to man from moose and roe deer in Sweden. Sci. Total Environ. 157, 309–316.

Johanson, K.J., Bergstrom, R., Eriksson, O., et al., 1994. Activity concentrations of Cs-137 in moose and their forage plants in mid-Sweden. J. Environ. Radioact. 22, 251–267.

Johnson, M.S., Roberts, R.D., 1978. Distribution of lead, zinc and cadmium in small mammals from polluted environments. OIKOS 30, 153–159.

Karasov, W.H., Jung, R.E., Van Den Langenberg, S., et al., 2005. Field exposure of frog embryos and tadpoles along a pollution gradient in the Fox River and Green River ecosystem in Wisconsin, USA. Environ. Toxicol. Chem. 24, 942–953.

Kashparov, V., Colle, C., Levchuk, S., et al., 2007. Radiochlorine concentration ratios for agricultural plants in various soil conditions. J. Environ. Radioact. 95, 10–22.

Kershaw, P.J., Mcmahon, C.A., Rudjord, A.L., et al., 2005. Spatial and temporal variations in concentration factors in NW European Seas – secondary use of monitoring data. Radioprotection 40 (Suppl. 1), S93–S99.

Khomich, V.K., 1990. On biological peculiarities of plants and absorption coefficients - Kn and Kd- for various agricultural crops after the Chernobyl accident, Radioecology of soil and plants. In: 3rd All-Union Conference on Agricultural Radiology, 2–7 July, 1990, vol 1, Russian Institute for Agricultural Radiology, Obninsk (in Russian).

Kulikov, N.V., Chebotina, M.YA., 1988. Radioecology of Fresh Water Biosystems. Nauka, Sverdlovsk (in Russian).

Kurabayashi, M., Fukuda, S., Kurokawa, Y., 1980. Concentration Factors of Marine Organisms Used for the Environmental Dose Assessment. In: Marine Radioecology (Proc. 3rd OECD/NEA Sem. Tokyo, 1979), OECD, Paris, p. 355.

Lee, D.M., 2006. Marine environmental radioactivity survey data. Aquat. Toxicol. 2.

Lindahl, P., Roos, P., Holm, E., et al., 2005. Studies of Np and Pu in the marine environment of Swedish–Danish waters and the North Atlantic Ocean. J. Environ. Radioact. 82, 285–301.

Livens, F.R., Horrill, A.D., Singleton, D.L., 1991. Distribution of radiocesium in the soil-plant systems of upland areas of Europe. Health Phys. 60, 539–544.

Mahon, D.C., Mathews, R.W., 1983. Uptake of naturally-occurring radioisotopes by vegetation in a region of high radioactivity. Can. J. Soil Sci. 63, 281–290.

Marshall, J.S., Wailerand, B.J., Yaguchi, E.M., 1975. Plutonium in the Laurentian great lakes: food-chain relationship. Verh. Internat. Verein. Limnol. 19, 323–329.

Mascanzoni, D., 1989a. Long-term transfer from soil to plant of radioactive corrosion products. Environ. Pollut. 57, 49–62.

Mascanzoni, D., 1989b. Plant uptake of activation and fission products in a long-term field study. J. Environ. Radioact. 10, 233–249.

Masson, M., Van Weers, A.W., Groothuis, R.E.J., et al., 1995. Time series for sea water and seaweed of Tc-99 and Sb-125 originating from releases at La Hague. J. Mar. Syst. 6, 397–413.

Matishov, G.G., Matishov, D.G., Szczypa, J., et al., 1994. Radionuclides in the Ecosystem of the Barents and Kara Seas Region. Kola Sci. Center Apatity (in Russian).

Matishov, G.G., Matishov, D.G., Namjatov, A.A., 1999. Modern level of the content of Cs-137 in fish and seaweed of the Barents Sea. In: Strand, P., Jølle, T. (Eds.), Extended Abstracts of the Fourth International Conference on Environmental Radioactivity in the Arctic, Edinburgh, Scotland, 20–23 September 1999. Norwegian Radiation Protection Authority, Østerås, Norway, pp. 242–243.

Melhuus, A., Seip, K.L., Seip, H.M., et al., 1978. A preliminary study of the use of benthic algae as biological indicators of heavy metal pollution in Sørfjorden, Norway. Environ. Pollut. 15, 101–107.

Mietelski, J.W., 2001. Plutonium in the environment of Poland (a review). In: Kudo, A. (Ed.), Plutonium in the Environment. Elsevier, Amsterdam, pp. 401–412.

Miretsky, G., Alekseev, P.V., Ramzaev, O.A., et al., 1993. New radioecological data for the Russian Federation (from Alaska to Norway). In: Strand, P., Holm, E. (Eds.), Environmental Radioactivity in the Arctic and Antarctic. Norwegian Radiation Protection Authority, Østerås, pp. 69–272.

Miroshichenko, T.A., Davydov, A.I., Us'Arov, A.G., 1990. Effects of genesis and physico-chemical properties of soils on 238U and 232Th accumulation by natural vegetation on uplands of the Central Caucasus radioecology of soil and plants. In: 3rd All-Union Conference on Agricultural Radiology, 2–7 July 1990, Obninsk, Vol. 1, Russian Institute of Agricultural Radiology and Agroecology, Obninsk, 80 (in Russian).

Mislevy, P., Blue, W.G., Roessler, C.E., 1989. Productivity of clay tailings from phosphate mining 1. Biomass Crops. J. Environ. Qual. 18, 95–100.

Morgan, J.E., Morgan, A.J., 1990. The distribution of cadmium, copper, lead, zinc and calcium in the tissues of the earthworm lumbricus-rubellus sampled from one uncontaminated and 4 polluted soils. Oecologia 84, 559–566.

Mortvedt, J.J., 1994. Plant and soil relationships of uranium and thorium decay series radionuclides – a review. J. Environ. Qual. 23, 643–650.

Mukhamedshin, L.D., Chilimov, B.I., Bezuglov, V.L., et al., 2000. Certification of forest resources by the radiation characteristic as the basis to obtain safe forest products in areas affected by radionuclides, In: Panchenko S.V. (Ed.), Problems of Forest Radioecology, MOGUL, Moscow, pp. 7–46 (in Russian).

Naustvoll, S., Ovrevoll, B., Hellstrom, T., et al., 1997. Radiocaesium in marine fish in the coastal waters of Northern Norway and in the Barents Sea. In: Strand, P. (Ed.), Extended Abstracts of the Third International Conference on Environmental Radioactivity in the Arctic. Tromsø, Norway, 1–5 June 1997. Norwegian Radiation Protection Authority, Østerås pp. 215–216.

Nelin, P., 1995. Radiocesium uptake in moose in relation to home-range and habitat composition. J. Environ. Radioact. 26, 189–203.

Nielsen, M.G., Gissel-Nielsen, G., 1975. Selenium in soil–animal relationships. Pedobiologia 15, 65–67.

NRPA, 1994. Radioactive Contamination at Dumping Sites for Nuclear Waste in the Kara Sea. Results from Norwegian–Russian 1993 Expedition to the Kara Sea. Norwegian Radiation Protection Authority, Østerås.

NRPA, 1995. Radioactivity in the Marine Environment. Sickel, M.A.K., Selnæs, T.D., Christensen, G.C., Bøe, B., Strand, P., Hellstrom, T. (Eds.). Norwegian Radiation Protection Authority, Østerås.

NRPA, 1997. Radioactive Contamination in the Marine Environment. In: Brungot, A.L., Sickel, M.A.K., Bergan, T.D., Boe, B., Hellstrom, T., Strand, P. (Eds.). Norwegian Radiation Protection Authority, Østerås.

NRPA, 1998. Technetium-99 Contamination in the North Sea and in Norwegian Coastal Areas 1996 and 1997. In: Brown J., Kolstad, A.K., Lind B., Rudjord, A.L., Strand P. (Eds.). Norwegian Radiation Protection Authority, Østerås.

NRPA, 1999. Radioactive Contamination in the Marine Environment. In: Brungot, A.L., Foen, L., Caroll, J., Kolstad, A.K., Brown, J., Rudjord, A.L., Bøe, B., Hellstrøm, T. (Eds.). Norwegian Radiation Protection Authority, Østerås.

NRPA, 2000. Radionuclide Uptake and Transfer in Pelagic Food Chains of the Barents Sea and Resulting Doses to Man and Biota. Final Report. Norwegian Transport and Effects Programme. Norwegian Radiation Protection Authority, Østerås.

Outola, I., Saxen, R., Heinävaara, S., 2009. Transfer of Sr-90 into fish in Finnish lakes. J. Environ. Radioact. 100, 657–664.

Pentreath, R.J., 1976. Monitoring of Radionuclides. FAO Fisheries Technical Paper 150. FAO, Rome, pp. 8–23.

Pentreath, R.J., 1981. The presence of ^{237}Np in the Irish Sea. Mar. Ecol. Prog. Ser. 6, 243–247.

Pentreath, R.J., Lovett, M.B., 1978. Transuranic nuclides in plaice (Pleuronectes platessa) from the north-eastern Irish Sea. Mar. Biol. 48, 19–26.

Perevolotsky, A.N., 2006a. ^{137}Cs and ^{90}Sr distribution in forest biogeocenoses. In: Perevolotsky, A.N. (Ed.), Institute for Radiology, Gomel, pp. 124–127 (in Russian).

Perevolotsky, A.N., 2006b. Effects of cultivation conditions on ^{137}Cs and ^{90}Sr accumulation levels in wood and bark of the major stand-forming species. In: Perevolotsky, A.N. (Ed.), Institute for Radiology, Gomel, pp. 267–273 (in Russian).

Pertsov, L.A., 1978. Biological Aspects of Radioactive Contamination of Sea. Atomizdat, Moscow (in Russian).

Pokarzhevskii, A., Zhulidov, A., 1995. Halogens in soil animal bodies: a background level. In: Van den Brink, W.J., Bosman, R., Arendt, F. (Eds.), Contaminated Soil. Academic Publishers, Dordrecht, pp. 403–404.

Pokarzhevskii, A.D., Krivolutzkii, D.A., 1997. Background concentrations of Ra-226 in terrestrial animals. Biogeochemistry 39, 1–13.

Polikarpov, G.G., 1964. Radioecology of Marine Organisms. Atomizdat, Moscow (in Russian).

Polikarpov, G.G., 1966. Radioecology of Aquatic Organisms: the Accumulation and Biological Effects of Radioactive Substances. Reinhold, New York.

Ponikarova, T.M., Drichko, V.F., Komarov, A.A., et al., 1990. Effects of hydrolysis lignin and surface-active substances on radiocaesium accumulation by plants. Radioecology of soil and plants. In: 3rd All-Union Conference on Agricultural Radiology, 2–7 July 1990, vol. 1, Russian Institute for Agricultural Radiology, Obninsk, pp. 87–88 (in Russian).

Preston, D.F., Dutton, J.W.R., 1967. The concentrations of caesium-137 and strontium-90 in the flesh of brown trout taken from rivers and lakes in the British Isles between 1961 and 1966. The variables determining the concentrations and their use in radiological assessments. Water Res. 1, 475–496.

Prister, B.S., Loshilov, N.A., Nemets, O.F., et al., 1988. Principles of agricultural radiology. In: Klimenko, R.F. (Ed.), Radionuclide Behaviour in the Soil–Plant System. Urozhay Press, Kiev, p. 163.

Read, J., Pickering, R., 1999. Ecological and toxicological effects of exposure to an acid, radioactive tailings storage. Environ. Monitor. Assess. 54, 69–85.

Rissanen, K., Matishov, D.G., Matishov, G.G., 1995. Radioactivity levels in Barents, Petshora, Kara Sea, Laptev and White Sea. In: International Conference on Environmental Radioactivity in the Arctic, 21–25 August. Oslo. Norwegian Radiation Protection Authority, Østerås, pp. 208–214.

Rissanen, K., Ikaheimonen, T.K., Matishov, D., et al., 1997. Radioactivity levels in fish, benthic fauna, seals, and sea birds collected in the northwest arctic of Russia. Radioprotection – Colloques 32, 323–331.

Rissanen, K., Ikaheimonen, T.K., Vlipieti, J., et al., 2000. Plutonium in algae, sediments and biota in the Barents, Pechora and Kara Seas. In: Inaba, J., Hisamatsu, S., Ohtsuka, Y. (Eds.), Proceedings of the International Workshop on Distribution and Speciation of Radionuclides in the Environment. Rokkasho, Aomori, pp. J107–J114.

Rowan, D.J., Rasmussen, J.B., 1994. Bioaccumulation of radiocesium by fish – the influence of physicochemical factors and trophic structure. Can. J. Fish. Aquat. Sci. 51, 2388–2410.

Rumble, M.A., Bjugstad, A.J., 1986. Uranium and radium concentrations in plants growing on uranium mill tailings in South-Dakota. Reclam. Reveget. Res. 4, 271–277.

Sanzharova, N.I., Abramova, T.N., Shukhovtsev, B.I., 1990. Sr-90 content in soils and agricultural products. Problems of agricultural radiology. Book of extended abstracts of the 3rd All-Union Conference on Agricultural Radiology, Vol. 4, Obninsk. 2–7 July, 1990, RIARAE, Obninsk, pp. 13–14.

Saxén, R., Outola, I., 2009. Polonium 210 in, freshwater and brackish environment. In: Gjelsvik R., Brown, J.E. (Eds.), A Deliverable Report for the NKS-B Activity October 2008, GAPRAD – Filling Knowledge Gaps in Radiation Protection Methodologies for Non-human Biota. Nordic Nuclear, Safety Research (NKS), DK - 4000 Roskilde, Denmark, pp. 13–21.

Sharmasarkar, S., Vance, G.F., 2002. Soil and plant selenium at a reclaimed uranium mine. J. Environ. Radioact. 31, 1516–1521.

Shcheglov, A.I., 1997. Biogeochemistry of Technogenic Radionuclides in Forest Ecosystems of the Central Regions of the East European Plain. Ph.D. Thesis, Moscow state University, Moscow.

Sheppard, S.C., Evenden, W.G., Macdonald, C.R., 1999. Variation among chlorine concentration ratios for native and agronomic plants. J. Environ. Radioact. 43, 65–76.

Shorti, Z.F., Palumbo, R.F., Oldon, P.B., et al., 1969. Uptake of I-131 by biota of Fern Lake, Washington in a laboratory and field experiment. Ecology 50, 979–989.

Shutov, V.N., Bekyasheva, T.A., Basalayeva, L.N., et al., 1993. Influence of soil properties on Cs-137 and Sr-90 radionuclides uptake by natural grasses. Pochvovedenie 8, 67–71.

Shutov, V.N., Bruk, G.YA., Travnikova, I.G., et al., 1999. The current radioactive contamination of the environment and foodstuffs in the Kola Region of Russia. In: Strand, P., Jølle, T. (Eds.), Extended Abstracts of the Fourth International Conference on Environmental Radioactivity in the Arctic, Edinburgh, Scotland, 20–23 September, 1999, Norwegian Radiation Protection Authority. Østerås, pp. 307–309.

Simon, S.L., Fraley, L., 1986. Uptake by sagebrush of uranium progeny injected in situ. J. Environ. Qual. 15, 345–350.

Stark, K., Avila, R., Wallberg, P., 2004. Estimation of radiation doses from Cs-137 to frogs in a wetland ecosystem. J. Environ. Radioact. 75, 1–14.

Steele, A.K., 1990. Derived concentration factors for Cs-137 in edible species of North Sea fish. Mar. Pollut. Bull. 21, 591–594.

Steinnes, E., Gaare, E., Engen, S., 2009. Influence of soil acidification in southern Norway on the Cs-137 exposure to moose? Sci. Total Environ. 407, 3905–3908.

Stephenson, M., Mierle, G., Reid, R.A., et al., 1994. Effects of experimental and cultural lake acidification on littoral benthic macroinvertebrate assemblages. Can. J. Fish. Aquat. Sci. 51, 1147–1161.

Tateda, Y., Koyanagi, T., 1996. Concentration factors for Cs-137 in Japanese coastal fish (1984–1990). J. Radiat. Res. 37, 71–79.

Templeton, W.L., 1959. Fission products and aquatic organisms. In: The Effects of Pollution on Living Material. Institute for Biology, London, pp. 125–140.

Titaeva, N.A., 1992. Nuclear geochemistry. In: Shchehura, I.I., Barinova, N.V. (Eds.), Uranium and Thorium Distribution in Plants. MSU Press, Moscow, pp. 78–80.

Titaeva, N.A., Toskaev, A.I., 1983. Uranium and thorium distribution in plants. In: Shchehura, I.I., Barinova, N.V. (Eds.), Nuclear Geochemistry. MSU Press, Moscow, pp. 54–58.

Tsvetnova, O.V., Sheglov, A.I., 2009. Cs-137 in components of natural ecosystems in the 30-km zone affected by Smolenskaya nuclear power plant. Pochvovedenie 17, 3–8.

Vakulovsky, S.M., 2008. The Radiation Situation Within Russia and Adjacent States in 2007. RosHydromet-SPA Typhoon, Obninsk (in Russian).

Van As, D., Fourie, H.O., Vleggaar, C.M., 1975. Trace element concentrations in marine organisms from the Cape West Coast. S.A.J. Sci. 71, 151–154.

Van Weers, Q.W., Van Raaphorst, J.G., 1979. Accumulation of trace elements in coastal marine organisms. Marine Radioecology. In: Proceedings 3rd OECD/NEA Sem. Tokyo, OECD, Paris (1980) p. 303.

Vandenhove, H., Van Hees, M., Wannijn, J., et al., 2006. Can we predict uranium bioavailability based on soil parameters? Part 2: Soil solution uranium concentration not a good bioavailability index. Environ. Pollut. 145, 577–586.

Vanderploeg, H.A., Parzcyk, D.C., Wilcox, W.H., et al., 1975. Bioaccumulation Factors for Radionuclides in Freshwater Biota. ORNL-5002. Oak Ridge National Laboratory, Oak Ridge, TN.

Verhovskaya, I.N., 1972. Radioecological Investigations in Natural Biogeocenoses. Nauka Moscow (in Russian).

Vermeulen, F., Van Den Brink, N.W., D'Havé, H., et al., 2009. Habitat type-based bioaccumulation and risk assessment of metal and As contamination in earthworms, beetles and woodlice. Environ. Pollut. 157, 3098–3105.

Vidal, M., Camps, M., Grebenshikova, N., et al., 2001. Soil- and plant-based countermeasures to reduce Cs-137 and Sr-90 uptake by grasses in natural meadows: the REDUP project. J. Environ. Radioact. 56, 139–156.

Wood, M.D., Marshall, W.A., Beresford, N.A., et al., 2008. Application of the ERICA integrated approach at the Drigg coastal sand dunes. J. Environ. Radioact. 99, 1484–1495.

Wood, M.D., Leah, R.T., Jones, S.R., et al., 2009. Radionuclide transfer to invertebrates and small mammals in a coastal sand dune ecosystem. Sci. Total Environ. 407, 4062–4074.

Wood, M.D., 2010. Assessing the Impact of Ionising Radiation in Temperate Coastal Sand Dunes: Measurement and Modelling. Ph.D. Thesis. University of Liverpool, Liverpool.

Yankovich, T.L., 2010. Compilation of Concentration Ratios for Aquatic Non-human Biota Collected by the Canadian Power Reactors Sector. CANDU Owners Group Inc., Toronto, Ontario, p. 7.

Yoshida, S., Muramatsu, Y., Peijnenburg, W.J.G.M., 2005. Multi-element analyses of earthworms for radioecology and ecotoxicology. Radioprotection 40, S491–S495.

ANNEX B. DERIVED CONCENTRATION RATIOS

Note that the wildlife groups noted in the tables (e.g. Grasses and Herbs, Shrub, Mammal, etc.) are categorised according to and use the same underlying data as IAEA (in preparation).

B.1. Terrestrial ecosystems

Table B.1. Wild Grass (*Poaceae*): derived concentration ratio (CR) values (units of Bq/kg fresh weight per Bq/kg dry weight for all elements except C, H, S, and P where units are Bq/kg fresh weight per Bq/m^3 air).

Element	Best estimate	Derivation method
Ag	1.8E+00	Assume CR for Grasses and Herbs; Ref ID: 162, 212
Ba	5.4E−02	Assume CR for Grasses and Herbs; Ref ID: 467, 518
C	8.9E+02	Specific activity model for Grasses and Herbs (Beresford et al., 2008a)
Ca	2.2E+00	Cereal stem and shoot CR from IAEA-472 (IAEA, 2010) assuming 25% dry matter
Ce	3.6E−03	Assume CR for Grasses and Herbs; Ref ID: 467
Cf	3.3E−02	Assume Pu CR for Wild Grass (Table A.1)
Cm	5.0E−04	Assume CR for Grasses; Ref ID: 491 (note $n = 1$ for this datum)
Co	3.9E−03	Assume CR for Grasses and Herbs; Ref ID: 467
Cr	5.8E−03	Assume CR for Grasses and Herbs; Ref ID: 467
Eu	3.6E−03	Assume CR for Grasses and Herbs; Ref ID: 467
H	1.5E+02	Specific activity model (Beresford et al., 2008a)
I	5.3E−02	Assume CR for Grasses; Ref ID: 179
Ir	4.0E−02	Assume Ru CR for Wild Grass (this table)
La	6.0E−03	Assume CR for Grasses and Herbs; Ref ID: 467
Mn	1.6E−01	Pasture CR from IAEA-472 (IAEA, 2010) assuming 25% dry matter
Nb	5.0E−03	Pasture CR from IAEA-472 (IAEA, 2010) assuming 25% dry matter
Np	1.5E−2	Pasture CR from IAEA-472 (IAEA, 2010) assuming 25% dry matter
P	8.9E+02	Assume CR for C (this table)
Pa	3.3E−02	Assume Pu CR for Wild Grass (Table A.1)
Ru	4.0E−2	Cereal stem and shoot CR from IAEA-472 (IAEA, 2010) assuming 25% dry matter
S	1.5E+02	CR given in Copplestone et al. (2003)
Te	2.5E−01	Pasture CR from IAEA-472 (IAEA, 2010) assuming 25% dry matter
Zr	2.5E−03	Pasture CR from IAEA-472 (IAEA, 2010) assuming 25% dry matter

Table B.2. Pine Tree (*Pinaceae*): derived concentration ratio (CR) values (units of Bq/kg fresh weight per Bq/kg dry weight for all elements except C, H, S, and P where units are Bq/kg fresh weight per Bq/m^3 air).

Element	Best estimate	Derivation method
Ag	1.9E–02	Assume CR for Shrub; Ref ID: 348
Am	1.7E–02	Assume CR for Shrub; Ref ID: 196, 486
C	1.3E+03	Specific activity model (Copplestone et al., 2001)
Ca	5.0E+00	Tagami and Uchida (2010); leguminous vegetable CR from IAEA-472 (IAEA, 2010) assuming 25% dry matter
Cd	3.5E–01	Assume CR for Tree; Ref ID: 180, 233
Cf	4.3E–02	Assume Pu CR for Shrub; Ref ID: 196, 468
Cm	9.4E–03	Assume CR for Tree – Broadleaf; Ref ID: 173
H	1.5E+02	Specific activity model for Grasses and Herbs (Beresford et al., 2008a)
I	5.3E–02	Assume CR for Grasses and Herbs; Ref ID: 179
Ir	3.2E–01	Assume Ru CR for Pine Tree (this table)
Mn	2.4E–02	Assume CR for Tree – Broadleaf; Ref ID: 255
Nb	5.0E–03	Pasture CR from IAEA-472 (IAEA, 2010)
Ni	1.8E–02	Assume CR for Tree – Broadleaf; Ref ID: 255
Np	4.3E–02	Assume Pu CR for Shrub; Ref ID: 196, 468
P	1.3E+03	Assume CR for C (this table)
Pa	4.3E–02	Assume Pu CR for Shrub; Ref ID: 196, 468
Pu	4.3E–02	Assume CR for Shrub; Ref ID: 196, 468
Ru	3.2E–01	Assume CR for Shrub; Ref ID: 468
S	1.5E+02	CR given in Copplestone et al. (2003)
Sb	3.2E+00	Assume CR for Shrub; Ref ID: 194, 467
Se	1.1E+00	Assume CR for Shrub; Ref ID: 248, 347, 348
Tc	8.4E–03	Assume CR for Shrub; Ref ID: 512
Te	2.5E–01	Pasture crop CR from IAEA-472 (IAEA, 2010)
Zr	7.2E–05	Assume CR for Shrub; Ref ID: 252

Table B.3. Earthworm (*Lumbricidae*): derived concentration ratio (CR) values (units of Bq/kg fresh weight per Bq/kg dry weight for all elements except C, H, S, and P where units are Bq/kg fresh weight per Bq/m^3 air).

Element	Best estimate	Derivation method
Ag	7.0E–01	Assume CR for Flying Insect from Beresford et al. (2008a)
Ba	3.8E–02	Assume CR for Arthropod; Ref ID: 518
C	4.3E+02	Specific activity model for Soil Invertebrate (Beresford et al., 2008a)
Ca	1.0E+01	Stable element data for Insecta from Bowen (1966)
Cf	1.1E+00	Assume Am CR for Earthworm (Table A.3)
Cm	1.1E+00	Assume Am CR for Earthworm (Table A.3)
Co	4.7E–03	Assume CR for Arthropod; Ref ID: 175, 234
Cr	5.0E–03	Assume CR from Bowen (1966)
H	1.5E+02	Specific activity model for Soil Invertebrate (Beresford et al., 2008a)
Ir	4.1E–03	Assume Ru CR for Earthworm (this table)
La	3.7E–04	Assume Ce CR for Earthworm (Table A.3)
Np	1.1E+00	Assume Am CR for Earthworm (Table A.3)
P	4.3+02	Assume CR for C (this table)
Pa	1.1E+00	Assume Am CR for Earthworm (Table 4.2)
Pu	2.1E–02	Assume CR for Annelid; Ref ID: 488,
Ra	2.1E+00	Assume CR for Arthropod; Ref ID: 192, 239, 388
Ru	4.1E–03	Assume CR for Arthropod; Ref ID: 175

Table B.3 (*continued*)

Element	Best estimate	Derivation method
S	5.0E+01	CR given in Copplestone et al. (2003)
Tc	3.5E−01	Assume Amphibian maximum animal value; Ref ID: 486
Te	3.8E−02	Assume Te CR for Gastropod from Beresford et al. (2008a)
Th	8.8E−03	Assume U CR for Annelid; Ref ID: 264
Zr	5.1E−04	Assume Nb CR for Earthworm (Table A.3)

Table B.4. Bee (*Apidea*): derived concentration ratio (CR) values (units of Bq/kg fresh weight per Bq/kg for all elements except C, H, S, and P where units are Bq/kg fresh weight per Bq/m^3 air).

Element	Best estimate	Derivation method
Ag	7.0E−01	Estimated from stable element data from Insects and Soils (Beresford et al., 2008a)
Am	4.0E−02	Assume CR for Arthropod; Ref ID: 170, 172, 223, 382, 407, 488
Ba	3.8E−02	Assume CR for Arthropod; Ref ID: 518
C	4.3E+02	Specific activity model based on Earthworm (Table B.3)
Ca	1.0E+01	Stable element data for Insects from Bowen (1966)
Cd	1.4E+00	Assume CR for Arthropod; Ref ID: 158, 202, 204, 254, 344
Ce	3.7E−04	Assume CR for Earthworm (Table A.3)
Cf	4.0E−02	Assume Am CR for Arthropod; Ref ID: 170, 172, 223, 382, 407, 488
Cl	2.8E−01	Assume CR for Arthropod; Ref ID: 238
Cm	1.4E−1	Assume Cm CR for Arthropod Ref ID: 223
Co	4.7E−03	Assume CR for Arthropod; Ref ID: 175, 234
Cr	5.0E−3	Assume CR for Flying Insect from Bowen (1966)
Cs	4.7E−3	Assume CR for Arthropod – Herbivorous; Ref ID: 170, 176
Eu	7.9E−04	Assume CR for Earthworm (Table A.3)
H	1.5E+02	Specific activity model based on Earthworm (Table B.3)
I	2.8E−01	Assume CR for Arthropod; Ref ID: 238
Ir	4.1E−03	Assume Ru CR for Bee (this table)
La	3.7E−04	Assume Ce CR for Earthworm (Table A.3)
Mn	4.4E−02	Assume CR for Mollusc; Ref ID: 191
Nb	5.1E−04	Assume CR for Earthworm (Table A.3)
Ni	8.6E−03	Assume CR for Arthropod; Ref ID: 234
Np	4.0E−02	Assume Am CR for Arthropod; Ref ID: 170, 172, 223, 382, 407, 488
P	4.3E+02	Assume C CR for Earthworm (this table)
Pa	4.0E−02	Assume Am CR for Arthropod; Ref ID: 170, 172, 223, 382, 407, 488
Pb	2.6E−01	Assume CR for Arthropod; Ref ID: 159, 204, 244, 344
Po	9.6E−02	Assume CR for Earthworm (Table A.3)
Pu	1.6E−02	Assume CR for Arthropod; Ref ID: 170, 216, 223, 261, 382, 407, 488
Ra	2.1E+00	Assume CR for Arthropod; Ref ID: 192, 239, 388
Ru	4.1E−03	Assume CR for Arthropod; Ref ID: 175
S	5.0E+01	CR given in Copplestone et al. (2003)
Sb	1.8E−01	Assume CR for Mollusc; Ref ID: 191
Se	1.5E+00	Assume CR for Earthworm (Table A.3)
Sr	8.4E−02	Assume CR for Arthropod; Ref ID: 169, 176, 223
Tc	3.5E−01	Assume amphibian maximum animal value; Ref ID: 486
Te	3.8E−02	Assume CR for Gastropod as reported in Beresford et al. (2008a)
Th	1.7E−02	Assume U CR for Arthropod (this table)
U	1.7E−02	Assume CR for Arthropod; Ref ID: 382
Zn	9.7E−01	Assume CR for Arthropod; Ref ID: 344
Zr	5.1E−04	Assume Nb CR for Earthworm (Table A.3)

Table B.5. Frog (*Ranidae*): derived concentration ratio (CR) values (units of Bq/kg fresh weight per Bq/kg for all elements except C, H, S, and P where units are Bq/kg fresh weight per Bq/m^3 air).

Element	Best estimate	Derivation method
Ag	2.9E−01	Assume CR value reported in Beresford et al. (2008a) and based upon stable element data presented for soil and humans
Ba	4.8E−03	Assume CR for Mammal; Ref ID: 518
C	1.3E+03	Specific activity model based on Mammal from Beresford et al. (2008a)
Ca	2.0E+00	Stable element review data (Bowen, 1979; Coughtrey and Thorne, 1983a,b) for soils and animals; based on mammalian data
Ce	6.1E−04	Allometric–biokinetic prediction for Mammal as reported in Beresford et al. (2008a)
Cf	1.0E−01	Assume Am CR for Frog (Table A.5)
Cl	7.0E+00	Assume CR reported in Beresford et al. (2008a) based on allometric–biokinetic model for Mammals
Cm	1.0E−01	Assume Am CR for Frog (Table A.5)
Co	1.8E−01	Assume CR for Mammal; Ref ID: 161
Cr	2.0E−04	Stable element review data (Bowen, 1979; Coughtrey and Thorne, 1983a,b) for soils and animals; based on mammalian data
Eu	2.0E−03	Allometric–biokinetic prediction for Mammal as reported in Beresford et al. (2008a)
H	1.5E+02	Specific activity model based on Mammal from Beresford et al. (2008a)
I	4.0E−01	Allometric–biokinetic prediction for Mammal as reported in Beresford et al. (2008a)
Ir	1.2E−01	Assume Ru CR for Frog (this table)
La	6.1E−04	Ce CR based on allometric–biokinetic prediction for Mammal as reported in Beresford et al. (2008a)
Mn	2.4E−03	Assume CR for Mammal; Ref ID: 199
Nb	1.90E−01	Assume CR reported in Beresford et al. (2008a). Estimated from stable element data presented for soil and (predominantly wild) animals in Coughtrey and Thorne (1983b)
Ni	3.0E−01	Assume CR for Reptile; Ref ID: 487
Np	1.0E−01	Assume Am CR for Frog (Table A.5)
P	1.3E+03	Assume C CR for Frog (this table)
Pa	1.0E−01	Assume Am CR for Frog (Table A.5)
Po	3.3E−02	Assume CR for Mammal; Ref ID: 61, 181, 182, 185, 186, 187, 196, 224, 225, 226, 227, 384, 423, 429, 450, 509
Pu	9.3E−03	Assume CR for Mammal; Ref ID: 172, 184, 197, 221, 222, 245, 261, 268, 405, 407, 488
Ra	1.7E−02	Assume CR for Mammal; Ref ID: 182, 185, 186, 187, 224, 225, 226, 227, 260, 423, 429, 458, 509
Ru	1.2E−01	Allometric–biokinetic prediction for Mammal as reported in Beresford et al. (2008a)
S	5.0E+01	CR given in Copplestone et al. (2003)
Sb	6.0E−02	Based on whole-body:diet CR with diet consisting of shrubs with shrub data from IAEA (in preparation)
Se	1.0E−02	Assume CR for Mammal; Ref ID: 246
Tc	3.5E−01	Assume CR for Amphibian; Ref ID: 486
Te	2.1E−01	Assume CR from Beresford et al. (2008a). Estimated from stable concentrations in soils and wild mammal tissues (Coughtrey et al., 1983)

Table B.5 (*continued*)

Element	Best estimate	Derivation method
Th	7.6E–02	Assume CR for Reptile; Ref ID: 450, 487
U	6.7E–01	Assume CR for Reptile; Ref ID: 450, 487
Zn	9.2E–02	Assume CR for Reptile; Ref ID: 487
Zr	1.2E–05	Assume CR reported in Beresford et al. (2008a) based on whole-body:diet CR with diet consisting of grass

Table B.6. Duck (*Anatidae*): derived concentration ratio (CR) values (units of Bq/kg fresh weight per Bq/kg for all elements except C, H, S, and P where units are Bq/kg fresh weight per Bq/Bq/m^3 air).

Element	Best estimate	Derivation method
Ag	2.9E–01	Assume CR reported in Beresford et al. (2008a) and based upon stable element data presented for soil and humans
Ba	4.8E–03	Assume CR for Mammal; Ref ID: 518
C	1.3E+03	Specific activity model for Bird: Beresford et al. (2008a)
Ca	2.0E+00	Stable element review data (Bowen, 1979; Coughtrey and Thorne, 1983a,b) for soils and animals; based on mammalian data
Cd	7.2E–01	Assume CR for Mammal; Ref ID: 158, 243
Ce	6.1E–04	Allometric–biokinetic prediction for Mammal as reported in Beresford et al. (2008a)
Cf	2.8E–02	Assume Am CR for Duck (Table A.6)
Cl	7.0E+00	Assume CR reported in Beresford et al. (2008a) based on allometric–biokinetic model for mammals
Cm	2.8E–02	Assume Am CR for Duck (Table A.6)
Co	1.8E–01	Assume CR for Mammal; Ref ID: 161
Cr	2.0E–04	Stable element review data (Bowen, 1979; Coughtrey and Thorne, 1983a,b) for soils and animals; based on mammalian data
Eu	2.0E–03	Allometric–biokinetic prediction for Mammal as reported in Beresford et al. (2008a)
H	1.5E+02	Specific activity model for Bird: Beresford et al. (2008a)
I	4.0E–01	Allometric–biokinetic prediction for Mammal as reported in Beresford et al. (2008a)
Ir	1.20E–01	Assume as Ru CR for Duck (this table)
La	6.1E–04	Ce CR for based on allometric–biokinetic prediction for Mammal as reported in Beresford et al. (2008a)
Mn	2.4E–03	Assume CR for Mammal; Ref ID: 199
Nb	1.90E–01	Assume CR reported in Beresford et al. (2008a). Estimated from stable element data presented for soil and (predominantly wild) animals in Coughtrey and Thorne (1983b)
Ni	3.1E–01	Assume CR for Reptile; Ref ID: 487
Np	2.8-02	Assume Am CR for Duck (Table A.6)
P	1.3E+03	Assume C CR for Duck (this table)
Pa	2.8E–02	Assume Am CR for Duck (Table A.6)
Pb	2.1E–02	Assume CR for Bird; Ref ID: 247
Po	9.6E–03	Assume CR for Bird; Ref ID: 384
Ru	1.20E–01	Allometric–biokinetic prediction for Mammal as reported in Beresford et al. (2008a)
S	5.0E+01	CR given in Copplestone et al. (2003)
Sb	6.0E–02	Based on whole-body:diet CR with diet consisting of shrubs with shrub data from IAEA (in preparation)

(*continued on next page*)

Table B.6 (*continued*)

Element	Best estimate	Derivation method
Se	1.0E–02	Assume CR for Mammal; Ref ID: 246
Te	2.1E–01	Assume CR from Beresford et al. (2008a). Estimated from stable concentrations in soils and wild mammal tissues from Coughtrey et al. (1983)
Th	3.8E–04	Assume CR for Bird; Ref ID: 260
U	4.9E–04	Assume CR for Bird; Ref ID: 260
Zn	9.2E–02	Assume CR for Reptile; Ref ID: 487
Zr	1.2E–05	Assume CR reported in Beresford et al. (2008a) based on whole-body:diet CR with diet consisting of grass

Table B.7. Rat (*Muridae*): derived concentration ratio (CR) values (units of Bq/kg fresh weight per Bq/kg for all elements except C, H, S, and P where units are Bq/kg fresh weight per Bq/m^3 air).

Element	Best estimate	Derivation method
Ag	2.9E–01	Assume CR reported in Beresford et al. (2008a) and based upon stable element data presented for soil and humans
Ba	4.8E–03	Assume CR for Mammal – Omnivorous; Ref ID: 518
C	1.3E+03	Specific activity model for Mammal (Beresford et al., 2008a)
Ca	2.0E+00	Stable element review data (Bowen, 1979; Coughtrey and Thorne, 1983a,b) for soils and animals; based on mammalian data
Cd	7.2E–01	Assume CR for Mammal; Ref ID: 158, 243
Ce	6.1E–04	Allometric–biokinetic prediction for Mammal as reported in Beresford et al. (2008a)
Cf	1.9E–2	Assume Pu CR for Rat (Table A.7)
Cl	7.0E+00	Assume CR reported in Beresford et al. (2008a) based on allometric–biokinetic model for Mammal
Cm	1.9E–2	Assume Pu CR for Rat (Table A.7)
Cr	2.0E–04	Stable element review data (Bowen, 1979; Coughtrey and Thorne, 1983a,b) for soils and animals; based on mammalian data
Eu	2.0E–03	Allometric–biokinetic prediction for Mammal as reported in Beresford et al. (2008a)
H	1.5E+02	Specific activity model for Mammal (Beresford et al., 2008a)
I	4.0E–01	Allometric–biokinetic prediction for Mammal as reported in Beresford et al. (2008a)
Ir	1.20E–01	Assume Ru CR for Rat (this table)
La	6.1E–04	Ce CR based on allometric–biokinetic prediction for Mammal as reported in Beresford et al. (2008a)
Mn	2.4E–03	Assume CR for Mammal; Ref ID: 199
Nb	1.90E–01	Assume CR reported in Beresford et al. (2008a). Estimated from stable element data presented for soil and (predominantly wild) animals in Coughtrey and Thorne (1983b)
Ni	7.2E–02	Assume CR for Mammal; Ref ID: 199
Np	1.9E–02	Assume Pu CR for Rat (Table A.7)
P	1.3E+03	Assume C CR for Rat (this table)
Pa	1.9E–02	Assume Pu CR for Rat (Table A.7)
Ru	1.20E–01	Allometric–biokinetic prediction for Mammal as reported in Beresford et al. (2008a)
S	5.0E+01	CR given in Copplestone et al. (2003)
Sb	6.0E–02	Based on whole-body:diet CR with diet consisting of shrubs with shrub data from IAEA (in preparation)

Table B.7 (*continued*)

Element	Best estimate	Derivation method
Se	1.0E−02	Assume CR for Mammal; Ref ID: 246
Tc	3.70E−01	Assume CR reported in Beresford et al. (2008a) based on allometric–biokinetic prediction
Te	2.1E−01	Assume CR from Beresford et al. (2008a). Estimated from stable concentrations in soils and wild mammal tissues from Coughtrey et al. (1983)
Zn	9.2E−02	Assume CR for Reptile; Ref ID: 487
Zr	1.2E−05	Assume CR reported in Beresford et al. (2008a) based on whole-body:diet CR with diet consisting of grass

Table B.8. Deer (*Cervidae*): derived concentration ratio (CR) values (units of Bq/kg fresh weight per Bq/kg for all elements except C, H, S, and P where units are Bq/kg fresh weight per Bq/Bq/m^3 air).

Element	Best estimate	Derivation method
Ag	2.9E−01	Assume CR reported in Beresford et al. (2008a) and based upon stable element data presented for soil and humans
Ba	4.8E−03	Assume CR for Mammal; Ref ID: 518
C	1.3E+03	Specific activity model for Mammal (Beresford et al., 2008a)
Ca	2.0E+00	Stable element review data (Bowen, 1979; Coughtrey and Thorne, 1983a,b) for soils and animals; based on mammalian data
Cd	6.7E+00	Assume CR for Mammal – Herbivorous; Ref ID: 158
Ce	6.1E−04	Allometric–biokinetic prediction for Mammal as reported in Beresford et al. (2008a)
Cf	2.1E−03	Assume Am CR for Deer (Table A.8)
Cl	7.0E+00	Assume CR reported in Beresford et al. (2008a) based on allometric–biokinetic model for Mammal
Cm	2.1E−03	Assume Am CR for Deer (Table A.8)
Co	1.8E−01	Assume CR for Mammal; Ref ID: 161
Cr	2.0E−04	Stable element review data (Bowen, 1979; Coughtrey and Thorne, 1983a,b) for soils and animals; based on mammalian data
Eu	2.0E−03	Allometric–biokinetic prediction for Mammal as reported in Beresford et al. (2008a)
H	1.5+02	Specific activity model for Mammal (Beresford et al., 2008a)
I	4.0E−01	Allometric–biokinetic prediction for Mammal as reported in Beresford et al. (2008a)
Ir	1.2E−01	Assume Ru CR for Deer (this table)
La	6.1E−04	Ce CR based on allometric–biokinetic prediction for Mammal (Beresford et al., 2008a)
Mn	2.4E−03	Assume CR for Mammal; Ref ID: 199
Nb	1.9E−01	Assume CR reported in Beresford et al. (2008a). Estimated from stable element data presented for soil and (predominantly wild) animals in Coughtrey and Thorne (1983b)
Ni	7.2E−02	Assume CR for Mammal; Ref ID: 199
Np	8.9E−04	Assume Pu CR for Deer (Table A.8)
P	1.3E+03	Assume C CR for Deer (this table)
Pa	8.9E−04	Assume Pu CR for Deer (Table A.8)
Pb	1.2E−02	Assume CR for Mammal – Herbivorous; Ref ID: 159, 181, 182, 185, 186, 187, 198, 211, 224, 225, 226, 227, 429

(*continued on next page*)

Table B.8 (*continued*)

Element	Best estimate	Derivation method
Po	2.4E−03	Assume CR for Mammal – Herbivorous; Ref ID: 181, 182, 185, 186, 187, 196, 224, 225, 226, 227, 429
Ra	6.1E−03	Assume CR for Mammal – Herbivorous; Ref ID: 182, 185, 186, 187, 224, 225, 226, 227, 260, 429
Ru	1.2E−01	Allometric–biokinetic prediction for Mammal as reported in Beresford et al. (2008a)
S	5.0E+01	CR given in Copplestone et al. (2003)
Sb	6.0E−02	Based on whole-body:diet CR with diet consisting of shrubs with shrub data from IAEA (in preparation)
Se	1.0E−02	Assume CR for Mammal; Ref ID: 246
Tc	3.7E−01	Assume CR reported in Beresford et al. (2008a) based on allometric–biokinetic prediction
Te	2.1E−01	Assume CR from Beresford et al. (2008a). Estimated from stable concentrations in soils and wild mammal tissues from Coughtrey et al. (1983)
Th	1.0E−04	Assume CR for Mammal – Herbivorous; Ref ID: 181, 182, 185, 186, 187, 224, 225, 226, 227
U	3.7E−03	Assume CR for Mammal; Ref ID: 61, 196, 423, 429, 450, 458, 509
Zn	9.2E−02	Assume CR for Reptile; Ref ID: 487
Zr	1.2E−05	Assume CR reported in Beresford et al. (2008a) based on whole-body:diet CR with diet consisting of grass

Table B.9. References for derived values for terrestrial Reference Animal and Plants (Tables B.1–B.8).

Ref ID	Reference	Ref ID	Reference
61	Williams (1981)	233	Opydo et al. (2005)
158	Andrews and Cooke (1982)	234	Peterson et al. (2003)
159	Andrews et al. (1989)	238	Pokarzhevskii and Zhulidov (1995)
161	Bastian and Jackson (1975)	239	Pokarzhevskii and Krivolutzkii (1997)
162	Beresford (1989)	243	Read and Martin (1993)
169	Cooper (2002)	244	Roberts et al. (1978)
170	Copplestone (1996)	245	Ryabokon et al. (2005)
172	Copplestone et al. (1999)	246	Sample and Suter (2002)
173	Coughtery et al. (1984)	247	Scheuhammer et al. (2003)
175	Crossley (1973)	248	Sharma and Shupe (1977)
176	Crossley (1961)	252	Sheppard and Evenden (1990)
179	Deitermann et al. (1989)	254	Skubala and Kafel (2004)
180	Efroymson et al. (2001)	255	Stanica (1999)
181	RIFE (2003)	260	Verhovskaya (1972)
182	RIFE (2004)	261	Whicker et al. (1974)
184	Ferenbaugh et al. (2002)	264	Yoshida et al. (2005)
185	RIFE (2000)	268	Beresford et al. (2008b)
186	RIFE (2001)	344	Vermeulen et al. (2009)
187	RIFE (2002)	347	Areva (2006)
191	Gaso et al. (2002)	348	Areva (2009)
192	Gaso et al. (2005)	382	Dragovic et al. (2010)
194	Ghuman et al. (1993)	384	Brown et al. (2009)
195	Gilhon (2001)	300	Dragovic and Jankovic Mandic (2010)
196	Green et al. (2002)	405	Gashchak and Beresford (2009)

Table B.9 (*continued*)

Ref ID	Reference	Ref ID	Reference
197	Hanson (1980)	407	Giles et al. (1990)
198	Haschek et al. (1979)	423	Lowson and Williams (1985)
199	Hendriks et al. (1995)	429	Martin et al. (1998)
202	Hunter and Johnson (1984)	450	Read and Pickering (1999)
204	Hussein et al. (2006)	458	Williams (1978)
211	Johnson and Roberts (1978)	467	Higley (2010)
212	Jones et al. (1985)	468	Panchenko and Panfilova (2000)
216	Little (1980)	486	Wood (2010)
221	Markham et al. (1978)	487	Wood et al. (2010)
222	Mietelski (2001)	488	Wood et al. (2009)
223	Mietelski et al. (2004)	491	Boone et al. (1981)
224	RIFE (1996)	509	Ryan et al. (2009)
225	RIFE (1997)	512	Tagami and Uchida (2005)
226	RIFE (1998)	518	Hope et al. (1996)
227	RIFE (1999)		

B.2. Freshwater ecosystems

Note that the wildlife groups noted in the tables (e.g. Freshwater Fish, Reptile etc.) are categorised according to and use the same underlying data as IAEA (in preparation).

Table B.10. Trout (*Salmonidae*): derived concentration ratio (CR) values (units of Bq/kg fresh weight per Bq/l).

Element	Best estimate	Derivation method
Ag	1.0E+02	Assume CR for Benthic Fish from Hosseini et al. (2008) based on published reviews
Am	5.7E+02	Assume CR for Freshwater Fish; Ref ID: 309, 411
Cd	1.9E+02	Assume CR for Freshwater Fish; Ref ID: 358, 391, 392, 427, 431, 441
Cf	2.0E+01	Assume Pu CR for Trout (Table A.10)
Cl	1.3E+02	Assume CR for Freshwater Fish; Ref ID: 304
Cm	5.7E+02	Assume Am CR for Freshwater Fish (this table)
H	1.0E+00	Simple specific activity assumption as reported in UNEP, ILO, WHO (1983).
Ir	2.9E+01	Assume Ru CR for Freshwater Fish (this table)
Nb	4.9E+02	Assume Zr CR for Trout (Table A.10)
Np	2.0E+01	Assume Pu CR for Trout (Table A.10)
Pa	2.0E+01	Assume Pu CR for Trout (Table A.10)
Ru	2.9E+01	Assume CR for Freshwater Fish; Ref ID: 301, 394
S	8.0E+02	Assume CR for Benthic Fish from Hosseini et al. (2008) based on published reviews
Tc	7.1E+01	Assume CR for Freshwater Fish; Ref ID: 301
Te	2.8E+02	Assume CR for Freshwater Fish; Ref ID: 333
Th	9.8E+01	Assume CR for Freshwater Fish; Ref ID: 304, 318, 339, 507

Table B.11. Frog (*Ranidae*): derived concentration ratio (CR) values (units of Bq/kg fresh weight per Bq/l).

Element	Best estimate	Derivation method
Ag	1.0E+02	Assume CR for Amphibian from Hosseini et al. (2008) based upon CR from a 'similar' organism (Benthic Fish) and based on published reviews
Am	5.7E+02	Assume CR for Freshwater Fish; Ref ID: 309, 411
Ba	4.3E+01	Assume CR for Freshwater Fish; Ref ID: 304, 333, 336, 339, 340, 343, 350, 355, 356, 357, 358, 359, 361, 363, 371, 376, 378, 517
C	7.3E+03	Assume CR from Hosseini et al. (2008). It is recommended that a specific activity approach be used to estimate C-14 in freshwater biota (Yankovich et al., 2008)
Cd	1.9E+02	Assume CR for Freshwater Fish; Ref ID: 358, 391, 392, 427, 431, 441
Ce	6.5E+01	Assume CR for Freshwater Fish; Ref ID: 304, 314, 333
Cf	3.8E+01	Assume Pu CR for Freshwater Fish; Ref ID: 301, 306, 307, 308, 309, 321, 331, 411, 462
Cl	1.3E+02	Assume CR for Freshwater Fish; Ref ID: 304
Cm	5.7E+02	Assume Am CR for Freshwater Fish (this table)
Co	8.2E+01	Assume CR for Freshwater Fish; Ref ID: 300, 301, 314, 324, 331, 333, 359, 394, 431, 445, 449, 461, 462, 517
Cs	1.6E+03	Assume CR for Freshwater Fish; Ref ID: 153, 178, 146, 300, 301, 302, 313, 314, 315, 319, 323, 326, 327, 331, 332, 333, 393, 394, 402, 408, 411, 415, 416, 418, 419, 445, 446, 454, 461, 462, 465
Eu	5.3E+01	Assume CR for Freshwater Fish; Ref ID: 304, 333
H	1E+00	Simple specific activity assumption as reported for Amphibian in Hosseini et al. (2008) and as reported in UNEP, ILO, WHO (1983)
I	2.6E+02	Assume CR for Freshwater Fish; Ref ID: 301, 314, 329, 333, 401, 517
Ir	2.9E+01	Assume Ru CR for Freshwater Fish (Table B.10)
La	6.0E+01	Assume CR for Freshwater Fish; Ref ID: 304, 333, 517
Mn	8.6E+02	Assume CR for Freshwater Fish; Ref ID: 314, 333, 336, 339, 340, 343, 350, 355, 356, 357, 358, 359, 361, 363, 364, 376, 378, 517
Nb	5.4E+01	Assume Zr CR for Freshwater Fish; Ref ID: 333, 517
Ni	1.2E+02	Assume CR for Freshwater Fish; Ref ID: 333, 336, 340, 343, 355, 356, 357, 358, 359, 374, 391, 441
Np	3.8E+01	Assume Pu CR for Freshwater Fish; Ref ID: 301, 306, 307, 308, 309, 321, 331, 411, 462
P	6.4E+05	Assume CR for Freshwater Fish; Ref ID: 333, 350
Pa	3.8E+01	Assume Pu CR for Freshwater Fish; Ref ID: 301, 306, 307, 308, 309, 321, 331, 411, 462
Po	5.9E+02	Assume CR for Freshwater Fish; Ref ID: 303, 311, 312, 328, 336, 339, 343, 346, 350, 355, 363, 383, 507
Pu	3.8E+01	Assume CR for Freshwater Fish; Ref ID: 301, 306, 307, 308, 309, 321, 331, 411, 462
Ra	5.5E+01	Assume CR for Freshwater Fish; Ref ID: 299, 301, 305, 318, 339, 340, 343, 346, 350, 355, 357, 361, 371, 507
Ru	2.9E+01	Assume CR for Freshwater Fish; Ref ID: 301, 394
S	8.0E+02	Assume CR for Amphibian from Hosseini et al. (2008) based upon CR from a 'similar' organism (Fish) and based on published reviews
Sb	1.1E+01	Assume CR for Freshwater Fish; Ref ID: 304, 333, 399
Se	4.0E+03	Assume CR for Freshwater Fish; Ref ID: 304, 310, 340, 356, 357, 359, 361, 371, 376, 378, 517
Sr	1.5E+02	Assume CR for Freshwater Fish; Ref ID: 178, 314, 317, 324, 331, 332, 333, 336, 339, 340, 350, 355, 356, 357, 358, 359, 361, 363, 371, 376, 389, 394, 411, 415, 416, 418, 419, 446, 454, 461, 462, 517

Table B.11 (*continued*)

Element	Best estimate	Derivation method
Tc	7.1E+01	Assume CR for Freshwater Fish; Ref ID: 301
Te	2.8E+02	Assume CR for Freshwater Fish; Ref ID: 333
Th	9.8E+01	Assume CR for Freshwater Fish; Ref ID: 304, 318, 339, 507
U	9.1E+00	Assume CR for Freshwater Fish; Ref ID: 299, 301, 303, 318, 339, 340, 350, 357, 358, 361, 371, 376, 377, 378, 507, 517
Zr	5.4E+01	Assume CR for Freshwater Fish; Ref ID: 333, 517

Table B.12. Duck (*Anatidae*): derived concentration ratio (CR) values (units of Bq/kg fresh weight per Bq/l).

Element	Best estimate	Derivation method
Ag	1.0E+02	Assume CR for Pelagic Fish from Hosseini et al. (2008)
Am	2.7E+02	Allometric–biokinetic model; fresh matter ingestion rate for 'All birds' from Nagy (2001); allometric loss rate from Brown et al. (2003); assume diet = vascular plant with geometric mean CR from Wildlife Transfer Database
Ba	3.9E+02	Assume Ca CR value (derived value based on Mammal) for Duck (this table)
C	7.3E+03	Assume CR from Hosseini et al. (2008) based upon 'highest available CR value' for organisms (Mollusc and Crustacean)
Ca	3.9E+02	Assume CR for Freshwater Mammal; Ref ID: 333
Cd	1.3E+03	Assume CR for Reptile; Ref ID: 487
Ce	6.3E+02	Assume CR for Reptile; Ref ID: 487
Cf	2.5E+02	Assume Pu CR (derived value) for Duck (this table)
Cl	8.2E+01	Assume CR from Hosseini et al. (2008) based on CR for a 'similar' organism (Fish)
Cm	2.7E+02	Assume Am CR (derived value) for Duck (this table)
Co	4.9E+02	Allometric–biokinetic model; fresh matter ingestion rate for 'All birds' from Nagy (2001); allometric loss rate from Higley et al. (2003); assume diet = vascular plant with geometric mean CR from Wildlife Transfer Database
Cr	6.8E+00	Allometric–biokinetic model; fresh matter ingestion rate for 'All birds' from Nagy (2001); allometric loss rate from RESRAD-Biota (Yu, 2007); assume diet = vascular plant with geometric mean CR from Wildlife Transfer Database
Cs	4.4E+02	Allometric–biokinetic model; fresh matter ingestion rate for 'All birds' from Nagy (2001); allometric loss rate from Brown et al. (2003); assume diet = vascular plant with geometric mean CR from Wildlife Transfer Database
Eu	5.0E+01	Assume CR from Hosseini et al. (2008) based on CR for a 'similar' organism (Fish)
H	1.0E+00	Simple specific activity model as reported for Bird in Hosseini et al. (2008) and as reported in UNEP, ILO, WHO (1983)
I	2.2E+02	Allometric–biokinetic model; fresh matter ingestion rate for 'All birds' from Nagy (2001); allometric loss rate from Brown et al. (2003); assume diet = vascular plant with geometric mean CR from Wildlife Transfer Database
Ir	2.0E+01	Assume Ru CR (derived value) for Duck (this table)

(*continued on next page*)

Table B.12 (*continued*)

Element	Best estimate	Derivation method
La	2.4E+02	Assume CR for Reptile; Ref ID: 487
Mn	1.5E+02	Assume CR for Freshwater Mammal; Ref ID: 511
Nb	2.3E+02	Assume CR from Hosseini et al. (2008) based upon CR for a 'similar' organism (Fish)
Ni	9.5E+02	Assume CR for Reptile; Ref ID: 487
Np	2.5E+02	Assume Pu CR (derived value) for Duck (this table)
P	6.2E+04	Assume CR from Hosseini et al. (2008) based on CR for a 'similar' organism (Benthic Fish)
Pa	2.5E+02	Assume Pu CR (derived value) for Duck (this table)
Pb	5.4E+00	Allometric–biokinetic model; fresh matter ingestion rate for 'All birds' from Nagy (2001); allometric loss rate from RESRAD-Biota (Yu, 2007); assume diet = vascular plant with geometric mean CR from Wildlife Transfer Database
Po	1.5E+02	Allometric–biokinetic model; fresh matter ingestion rate for 'All birds' from Nagy (2001); allometric loss rate from RESRAD-Biota (Yu, 2007); assume diet = vascular plant with geometric mean CR from Wildlife Transfer Database
Pu	2.5E+02	Allometric–biokinetic model; fresh matter ingestion rate for 'All birds' from Nagy (2001); allometric loss rate from Brown et al. (2003); assume diet = vascular plant with geometric mean CR from Wildlife Transfer Database
Ra	3.7E+02	Assume CR for Reptile, geometric mean; Ref ID: 487
Ru	2.0E+01	Assume CR from Hosseini et al. (2008) based upon CR for a 'similar' organism (Fish)
S	8.0E+02	Assume CR from Hosseini et al. (2008) based upon CR for a 'similar' organism (Fish)
Sb	2.3E+03	Assume CR for Reptile; Ref ID: 487
Se	1.9E+03	Assume CR for Reptile; Ref ID: 487
Sr	4.1E+03	Allometric–biokinetic model; fresh matter ingestion rate for 'All birds' from Nagy (2001); allometric loss rate from Brown et al. (2003); assume diet = vascular plant with geometric mean CR from Wildlife Transfer Database
Tc	4.0E+01	Assume CR from Hosseini et al. (2008) based upon CR for a 'similar' organism (Fish)
Te	7.0E+02	Assume CR from Hosseini et al. (2008) based upon CR for a 'similar' organism (Fish)
Th	2.8E+03	Allometric–biokinetic model; fresh matter ingestion rate for 'All birds' from Nagy (2001); allometric loss rate from Brown et al. (2003); assume diet = vascular plant with geometric mean CR from Wildlife Transfer Database
U	1.3E+01	Allometric–biokinetic model; fresh matter ingestion rate for 'All birds' from Nagy (2001); allometric loss rate from Brown et al. (2003); assume diet = vascular plant with geometric mean CR from Wildlife Transfer Database
Zn	1.2E+04	Allometric–biokinetic model; fresh matter ingestion rate for 'All birds' from Nagy (2001); allometric loss rate from RESRAD-Biota (Yu, 2007); assume diet = vascular plant with geometric mean CR from Wildlife Transfer Database
Zr	1.2E+03	Assume CR for Reptile; Ref ID: 487

Table B.13. References for derived values for freshwater Reference Animals and Plants (Tables B.10–B.12).

Ref ID	Reference	Ref ID	Reference
146	Vakulovsky (2008)	357	Cameco (2007)
153	Vintsukevich and Tomilin (1987)	358	Cameco (2009a)
178	Vetikko and Saxen (2010)	359	Cameco (2009b)
299	BEAK (1987)	361	Cameco (2001)
300	Bird (1998)	363	Cameco (2005a)
301	Blaylock (1982)	364	Cameco (2008c)
302	Carlsson and Lidén (1978)	371	Cameco (2000)
303	Carvalho et al. (2007)	374	Cameco (2005b)
304	Chapman et al. (1968)	376	Cameco (2005c)
305	Clulow et al. (1998)	377	Cameco (2003b)
306	Edgington et al. (1976)	378	Cameco (2005d)
307	Emery et al. (1976)	383	Saxen and Outola (2009)
308	Eyman and Trabalka (1980)	389	Outola et al. (2009)
309	Garten et al. (1981)	391	Ahmad et al. (2010)
310	Graham et al. (1992)	392	Al-Kahtani (2009)
311	Hameed et al. (1993)	393	Antonenko (1978)
312	Hameed et al. (1997)	394	Apostoaer et al. (1999)
313	Hewett and Jefferies (1978)	399	Culioli et al. (2009)
314	Jinks and Eisenbud (1972)	401	Dubynin (1987)
315	Kevern and Spigarelli (1971)	402	Dushauskene-Duzh (1969)
317	Krumholz (1956)	408	Golubev et al. (2007)
318	Lambrechts et al. (1992)	411	Gudkov et al. (2005)
319	Linder et al. (1990)	415	Kryshev and Ryabov (2005)
321	Marshall et al. (1975)	416	Kulikov and Chebotina (1988)
323	Newman and Brisbin (1990)	418	Kulikov and Kulikova (1977)
324	Ophel et al. (1972)	419	Kulikov and Molchanova (1975)
326	Preston and Dutton (1967)	427	Malik et al. (2010)
327	Rowan and Rasmussen (1994)	431	Mohamed (2008)
328	Shaheed et al. (1997)	441	Ozturk et al. (2009)
329	Shorti et al. (1969)	445	Trapeznikov (2001)
331	Trapeznikov et al. (1993a)	446	Zesenko and Kulebyakina (1982)
332	Vanderploeg et al. (1975)	449	Rashed (2001)
333	Yankovich (2010)	454	Smagin (2006)
336	Areva (2010)	461	Trapeznikov et al. (2007)
339	COGEMA (2005)	462	Trapeznikov et al. (1993b)
340	Areva (2007)	465	Trapeznikova et al. (1984)
343	COGEMA (1998)	487	Wood et al. (2010)
346	Areva (2006)	507	Martin et al. (1995)
350	Cameco (2003a)	511	Golder Associates Ltd. (2005)
355	Cameco (2008a)	517	Engdahl et al. (2006)
356	Cameco (2008b)		

B.3. Marine ecosystems – derived concentration ratio values

Note that the wildlife groups noted in the tables (e.g. Macro-algae, Marine Crustacean, etc.) are categorised according to and use the same underlying data as IAEA (in preparation).

Table B.14. Brown Seaweed (*Fucaceae*): derived concentration ratio values (units of Bq/kg fresh weight per Bq/l).

Element	Best estimate	Derivation method
Ba	1.6E+03	Assume CR for Estuarine Macro-algae; Ref ID: 506, 517
C	8.0E+03	Assume CR for Marine Macro-algae; Ref ID: 21
Ca	3.8E+00	Assume CR for Estuarine Macro-algae; Ref ID: 101, 439, 506, 517
Cf	7.7E+01	Assume Am CR for Brown Seaweed (Table A.14)
Cl	7.3E−01	Assume CR for Marine Macro-algae; Ref ID: 21, 65
Cr	3.5E+02	Assume CR for Estuarine Macro-algae; Ref ID: 506
Eu	1.1E+03	Assume CR for Marine Macro-algae; Ref ID: 141
I	1.4E+03	Assume CR for Marine Macro-algae; Ref ID: 10, 21, 62, 65, 120
Ir	1.0E+03	Recommended CR for Macro-algae from IAEA (2004)
La	5.9E+03	Assume CR for Estuarine Macro-algae; Ref ID: 101, 439
P	9.6E+03	Assume CR for Marine Macro-algae; Ref ID: 21
Pa	1.0E+02	Recommended CR for Marine Macro-algae from IAEA (2004)
Po	7.1E+02	Assume CR for Marine Macro-algae; Ref ID: 4, 28, 29, 46, 95, 133
Ra	4.4E+01	Assume CR for Marine Macro-algae; Ref ID: 18, 29
S	2.4E+00	Assume CR for Marine Macro-algae; Ref ID: 21
Se	2.0E+02	Assume CR for Marine Macro-algae; Ref ID: 65, 87
Te	1.0E+04	Recommended CR for Macro-algae from IAEA (2004)
Th	2.4E+03	Assume CR for Marine Macro-algae; Ref ID: 29, 64, 100
Zn	1.3E+04	Assume CR for Estuarine Macro-algae; Ref ID: 506

Table B.15. Crab (*Cancridae*): derived concentration ratio values (units of Bq/kg fresh weight per Bq/l).

Element	Best estimate	Derivation method
Ag	2.0E+05	Recommended CR for Crustaceans from IAEA (2004)
Am	5.0E+02	Assume CR for Marine Crustacean; Ref ID: 133
Ba	8.0E+02	Assume CR for Estuarine Crustacean; Ref ID: 506
C	1.0E+4	Assume CR for Marine Crustacean; Ref ID: 21
Ca	4.5E+00	Assume CR for Estuarine Large Crustacean; Ref ID: 101, 439
Ce	1.0E+2	Assume CR for Marine Crustacean; Ref ID: 83
Cf	5.0E+02	Assume Am CR for Marine Crustacean (this table)
Cl	5.6E−02	Assume CR for Marine Crustacean; Ref ID: 21
Cm	5.0E+02	Assume Am CR for Marine Crustacean (this table)
Co	4.7E+03	Assume CR for Marine Large Crustacean; Ref ID: 120, 147
Cr	2.8E+02	Assume CR for Estuarine Crustacean; Ref ID: 506
Eu	2.4E+04	Assume CR for Estuarine Large Crustacean; Ref ID: 101, 439
H	1.0E+00	Recommended CR for Crustaceans from IAEA (2004) – tritiated water
I	3.0E+00	Recommended CR for Crustaceans from IAEA (2004)
Ir	1.0E+02	Recommended CR for Crustaceans from IAEA (2004)
La	1.0E+02	Assume Ce CR for Marine Crustacean (this table)
Mn	2.5E+03	Assume CR for Marine Large Crustacean; Ref ID: 53, 85
Nb	1.0E+02	Assume CR for Marine Crustacean; Ref ID: 10
Ni	9.1E+02	Assume CR for Estuarine Large Crustacean; Ref ID: 101, 439

Table B.15 (*continued*)

Element	Best estimate	Derivation method
Np	1.1E+02	Assume CR for Marine Large Crustacean; Ref ID: 515
P	3.0E+04	CR derived from stable P in Crustaceans from Hosseini et al. (2008)
Pa	1.0E+01	Recommended CR for Crustaceans from IAEA (2004)
Pb	3.4E+03	Assume CR for Marine Large Crustacean; Ref ID: 4, 59
Po	4.2E+03	Assume CR for Marine Large Crustacean; Ref ID: 4
Ra	7.3E+01	Assume CR for Crustacean; Ref ID: 96
Ru	1.0E+02	Recommended CR for Crustaceans from IAEA (2004)
Sb	3.0E+02	Recommended CR for Crustaceans from IAEA (2004)
Se	1.0E+04	Recommended CR for Crustaceans from IAEA (2004)
Te	1.0E+03	Recommended CR for Crustaceans from IAEA (2004)
Th	1.0E+03	Recommended CR for Crustaceans from IAEA (2004)
U	6.2E+00	Assume CR for Estuarine Large Crustacean; Ref ID: 439
Zn	3.0E+05	Recommended CR for Crustaceans from IAEA (2004)
Zr	4.9E+01	Assume CR for Marine Crustacean; Ref ID: 83

Table B.16. Flatfish (*Pleuronectidae*): derived concentration ratio values (units of Bq/kg fresh weight per Bq/l).

Element	Best estimate	Derivation method
Ag	8.1E+03	Assume CR for Marine Fish; Ref ID: 8, 21, 31
Ba	9.6E+00	Assume CR for Estuarine Fish, Ref ID: 506, 517
C	1.2E+04	Assume CR for Marine Fish; Ref ID: 21
Cd	1.3E+04	Assume CR for Marine Fish; Ref ID: 10, 31, 36, 87
Ce	2.1E+02	Assume CR for Marine Fish; Ref ID: 83, 141
Cf	1.9E+02	Assume Am CR for Flatfish (Table A.16)
Cl	6.2E−02	Assume CR for Estuarine Fish, Ref ID: 506
Cm	1.9E+02	Assume Am CR for Flatfish (Table A.16)
Cr	2.0E+02	Recommended CR for Fish from IAEA (2004)
Eu	7.3E+02	Assume CR for Marine Fish; Ref ID: 141
H	1.0E+00	Recommended CR for Fish from IAEA (2004) – tritiated water
I	9.0E+00	Recommended CR for Fish from IAEA (2004)
Ir	2.0E+01	Recommended CR for Fish from IAEA (2004)
La	2.1E+02	Assume Ce CR for Marine Fish; Ref ID: 83, 141
Nb	3.0E+01	Recommended CR for Fish from IAEA (2004)
Np	2.1E+01	Assume Pu CR for Flatfish (Table A.16)
P	9.5E+04	Assume CR for Marine Fish; Ref ID: 21, 75
Pa	5.0E+01	Recommended CR for Fish from IAEA (2004)
Po	1.2E+04	Assume CR for Marine Fish; Ref ID: 4, 28, 29, 46, 51
Ra	6.3E+01	Assume CR for Fish – Benthic Feeding; Ref ID: 96, 121
Ru	1.6E+01	Assume CR for Marine Fish; Ref ID: 10
S	1.0E+00	Recommended CR for Fish from IAEA (2004)
Sb	6.0E+02	Recommended CR for Fish from IAEA (2004)
Se	1.0E+04	Recommended CR for Fish from IAEA (2004)
Tc	8.0E+01	Recommended CR for Fish from IAEA (2004)
Te	1.0E+03	Recommended CR for Fish from IAEA (2004)
Th	1.3E+03	Assume CR for Marine Fish; Ref ID: 29
U	4.0E+00	Assume CR for Fish – Benthic Feeding; Ref ID: 122

Table B.17. References for derived values for marine Reference Animal and Plants (Tables B.14–B.16).

Ref ID	Reference	Ref ID	Reference
4	Al-Masri et al. (2000)	83	Kurabayashi et al. (1980)
8	Amiard (1978)	85	Lentsch et al. (1973)
10	Ancellin et al. (1979)	87	Locatelli and Torsi (2000)
18	Bonotto et al. (1981)	95	Mcdonald et al. (1992)
21	Bowen (1979)	96	Meinhold and Hamilton (1992)
28	Carvalho (1988)	100	Nilsson et al. (1981)
29	Cherry and Shannon (1974)	101	Takata et al. (2010)
31	Cohen (1985)	120	Polikarpov (1966)
36	Coughtrey and Thorne (1983c)	121	Porntepkasemsan and Nevissi (1990)
46	Folsom et al. (1973)	122	Poston and Klopfer (1986)
51	Gomez et al. (1991)	133	Sivintsev et al. (2005)
53	Guthrie et al. (1979)	141	Suzuki et al. (1975)
59	Heyraud and Cherry (1979)	147	Van As et al. (1975)
62	Holm et al. (1994)	439	NIRS (2009)
64	Holm and Persson (1980)	506	Kumblad and Bradshaw (2008)
65	Hou and Yan (1998)	515	Pentreath (1981)
75	Kahn (1980)	517	Engdahl et al. (2006)

References

Ahmad, M.K., Islam, S., Rahman, S., et al., 2010. Heavy metals in water, sediment and some fishes of Buriganga River, Bangladesh. Int. J. Environ. Res. 4, 321–332.

Al-Kahtani, M.A., 2009. Accumulation of heavy metals in talapia fish (Oreochromis niloticus) from Al-Khadoud spring, Al-Hassa, Saudi Arabia. Am. J. Appl. Sci. 6, 2024–2029.

Al-Masri, M.S., Mamish, S., Budeir, Y., 2000. Assessment of the potential impact of the phosphate industry along the Syrian coast by evaluating Po-210 and Pb-210 levels in sediment, seawater and selected marine organisms. AECS-PR\FRSR 232 Atomic Energy Commission of Syria, Damascus, 27 pp.

Amiard, J.C., 1978. A Study of the Uptake and Toxicity of Some Stable and Radioactive Pollutants in Marine Organisms: Antimony, Silver, Cobalt and Strontium in Mollusks, Crustaceans and Teleosts. Commissariat a l'Energie Atomique (CEA)-R4928, CEA, Saclay, France.

Ancellin, J., Gueguеniat, P., Germaine, P., 1979. Radioecologie Marine. Eyrolles, Paris.

Andrews, S.M., Cooke, J.A., 1982. Cadmium within a contaminated grassland ecosystem established on metalliferous mine waste. In: Osbourn, D. (Ed.), Metals in Animals. CEH Monkswood, Huntingdon.

Andrews, S.M., Johnson, M.S., Cooke, J.A., 1989. Distribution of trace-element pollutants in a contaminated grassland ecosystem established on metalliferous fluorspar tailings. 1. Lead. Environ. Pollut. 58, 73–85.

Antonenko, T.M., 1978. Radioecological study of Cs-137 accumulation, distribution and migration in the water bodies of the steppe zone of the Ukraine. PhD Thesis INBUM Sevastopol (1978) (in Russian).

Apostoaer, A.I., Blaylock, B.G., Caldwell, B., et al., 1999. Radionuclide releases from X-10 to the Clinch River – Measurements in the flesh of edible species of fish. Task 4 Report. Radionuclide release to the Clinch river from White Oak Creek on the Oak Ridge Reservation – an Assessment of Historical Quantities ChemRisk/SENES Oak Ridge Inc., Oak Ridge Tennessee.

Areva, 2006. McClean Lake Operation Status of the Environment Report. Assessment Period 2003–2005, Saskatoon, Saskatchewan.

Areva, 2007. Pore-water Study Using In Situ Dialysis for the Link Lakes at the Rabbit Lake Operation. Canada North Environmental Services, Saskatoon, Saskatchewan.
Areva, 2009. McClean Lake Operation Status of the Environment Report. Assessment Period 2006–2008, Saskatoon, Saskatchewan.
Areva, 2010. Shea Creek Project Area, Environmental Baseline Investigation 2007–2009. Draft Report. Canada North Environmental Services, Saskatoon, Saskatchewan.
Bastian, R.K., Jackson, W.B., 1975. Cs-137 and Co-60 in a terrestrial community at Enewatak Atoll. In: Cushing, C.E.J. (Ed.), Proceedings of a Symposium on Radioecology and Energy Resources. Academic Press, London, pp. 313–320.
BEAK, 1987. Survey of Data on the Radionuclide Content of Fish in Canada. Report prepared for the Atomic Energy Control Board by Beak Consultants Ltd. Atomic Energy Control Board, Ottawa, Canada.
Beresford, N.A., 1989. Estimating the transfer of 110mAg originating from the Chernobyl accident in west Cumbrian soil and vegetation samples. J. Radiol. Prot. 9, 281–283.
Beresford, N.A., Barnett, C.L., Howard, B.J., et al., 2008a. Derivation of transfer parameters for use within the ERICA Tool and the default concentration ratios for terrestrial biota. J. Environ. Radioact. 99, 1393–1407.
Beresford, N.A., Gaschak, S., Barnett, C.L., et al., 2008b. Estimating the exposure of small mammals at three sites within the Chernobyl exclusion zone – a test application of the ERICA Tool. J. Environ. Radioact. 99, 1496–1502.
Bird, G., 1998. Fate of Co-60 and Cs-134 added to the hypolimnion of a Canadian Shield Lake: Accumulation in biota. Can. J. Fish. Aqu. Sci. 55, 987–998.
Blaylock, B.G., 1982. Radionuclide data bases available for bioaccumulation factors for freshwater biota. Nucl. Saf. 23, 427–438.
Bonotto, S., Carraro, G., Stract, S., et al., 1981. Ten years of investigations on radioactive contamination of the marine environment. In: Impacts of Radionuclide Releases into the Marine Environment, Proceedings Symposium, 6–10 October 1980, Vienna, 1980. IAEA-SM-248/105, pp. 649–660.
Boone, F.W., Ng, C., Palms, J.M., 1981. Terrestrial pathways of radionuclide particulates. Health Phys. 41, 735–747.
Bowen, H.J.M., 1966. Trace Elements in Biochemistry. Academic Press, London.
Bowen, H.J.M., 1979. Environmental Chemistry of the Elements. Academic Press, London.
Brown, J., Gjelsvik, R., Holm, E., et al., 2009. Filling knowledge gaps in radiation protection methodologies for non-human biota. Final summary report. Nordic Nuclear Safety Research (NKS) Report. ISBN 978-87-7893-254-9, NKS, DK – 4000, Roskilde, Denmark, p. 17.
Brown, J.E., Strand, P., Hosseini, A., Børretzen, P. (Eds.), 2003. Handbook for Assessment of the Exposure of Biota to Ionising Radiation from Radionuclides in the Environment. Norwegian Radiation Protection Authority, Østerås.
Cameco, 2000. Current Period Environmental Monitoring Program for the Beaverlodge Mine Site – Revision 2. Conor Pacific Environmental Technologies Inc., Saskatoon, Saskatchewan.
Cameco, 2001. Technical Memorandum – Water Quality Results from Eagle Drill Hole and Dubyna Drill Hole. Canada North Environmental Services, Saskatoon, Saskatchewan.
Cameco, 2003a. Beaverlodge Decommissioning, Results of the 2002 Aquatic Biological Investigations at the Dubyna Mine Site Area, Northern Saskatchewan. Canada North Environmental Services, Saskatoon, Saskatchewan.
Cameco, 2003b. Rabbit Lake Environmental Effect Monitoring/Environmental Monitoring Program, 2002. Golder Associates, Saskatoon, Saskatchewan.
Cameco, 2005a. The Cigar Lake Uranium Project, Environmental Effects Monitoring and Biological Monitoring Studies, 2004. Interpretive Report. Canada North Environmental Services, Saskatoon, Saskatchewan.
Cameco, 2005b. Key Lake Operation Comprehensive Environmental Effects Monitoring Program. Interpretive Report. Golder Associates, Saskatoon, Saskatchewan.
Cameco, 2005c. McArthur River Operation Comprehensive Environmental Effects Monitoring Program. Interpretative Report. Golder Associates, Saskatoon, Saskatchewan.

Cameco, 2005d. Rabbit Lake Uranium Operation Comprehensive Environmental Effect Monitoring Program. Interpretive Report. Golder Associates, Saskatoon, Saskatchewan.
Cameco, 2007. Pore-water study Using In Situ Dialysis for the Link Lakes at the Rabbit Lake Operation. Canada North Environmental Services, Saskatoon, Saskatchewan.
Cameco, 2008a. Biophysical Baseline Program for the Millennium Project Area. Draft Report. Canada North Environmental Services, Saskatoon, Saskatchewan.
Cameco, 2008b. Cigar Lake Project, 2007. Comprehensive Aquatic Environment Monitoring Report. Canada North Environmental Services, Saskatoon, Saskatchewan.
Cameco, 2008c. McArthur River Operation, 2007. Comprehensive Aquatic Environment Monitoring Report. Canada North Environmental Services, Saskatoon, Saskatchewan.
Cameco, 2009a. Rabbit Lake Operation, 2008. Comprehensive Aquatic Environment Monitoring Report. Canada North Environmental Services, Saskatoon, Saskatchewan.
Cameco, 2009b. Results of the 2009 Key Lake Northern Pike Chemistry Monitoring Program. Canada North Environmental Services, Saskatoon, Saskatchewan.
Carlsson, S., Lidén, K., 1978. ^{137}Cs and potassium in fish and littoral plants from a humus-rich oligotrophic lake 1961–1976. Oikos 30, 126–132.
Carvalho, F.P., 1988. Po-210 in marine organisms: a wide range of natural radiation dose domains. Radiat. Prot. Dosim. 24, 113–117.
Carvalho, F.P., Oliveira, J.M., Lopes, I., et al., 2007. Radionuclides from past uranium mining in rivers of Portugal. J. Environ. Radioact. 98, 298–314.
Chapman, W.H., Fisher, H.L., Pratt, M.W., 1968. Concentration Factors of Chemical Elements in Edible Aquatic Organisms. Report No. UCRL-50564. National Technical Information Service, Springfield, VA.
Cherry, R.D., Shannon, L.V., 1974. The alpha radioactivity of marine organisms. Atom. Energy Rev. 12, 3–45.
Clulow, F.V., Dave, N.K., Lim, T.P., et al., 1998. Radium-226 in water, sediments, and fish from lakes near the city of Elliot Lake, Ontario, Canada. Environ. Pollut. 99, 13–28.
COGEMA, 1998. Cluff Lake Project, Suspension of Operations and Eventual Decommissioning of the Tailings Management Area, TMA, Biological Environment. Conor Pacific Environmental Technologies Inc., Saskatoon, Saskatchewan.
COGEMA, 2005. Cluff Lake Uranium Mine 2004. Environmental Effects Monitoring and Environmental Monitoring Programs. Canada North Environmental Services, Saskatoon, Saskatchewan.
Cohen, B.L., 1985. Bioaccumulation factors in marine organisms. Health Phys. 49, 1290–1294.
Cooper, K., 2002. The Effect of Chronic Radiation on Invertebrate Diversity and Abundance Within the Chernobyl Exclusion Zone. M.Sc. Thesis (Restoration Ecology). University of Liverpool, Liverpool, p. 62.
Copplestone D., Bielby S., Jones S.R., et al., 2001. Impact Assessment of Ionising Radiation on Wildlife. R&D Publication 128. Environment Agency, Bristol.
Copplestone, D., 1996. Chapter 3. A Coniferous Woodland Ecosystem – Lady Wood. The Food Chain Transfer of Radionuclides Through Semi Natural Habitats. Ph.D. Thesis. University of Liverpool, Liverpool, pp. 77–164.
Copplestone, D., Johnson, M.S., Jones, S.R., et al., 1999. Radionuclide behaviour and transport in a coniferous woodland ecosystem: vegetation, invertebrates and wood mice, Apodemus sylvaticus. Sci. Total Environ. 239, 96–109.
Copplestone, D., Wood, M.D., Bielby, S., et al., 2003. Habitat Regulations for Stage 3 Assessments: Radioactive Substances Authorisations. R&D Technical Report P3-101/SP1a. Environment Agency, Bristol.
Coughtrey, P.J., Thorne, M.C., 1983a. Radionuclide Distribution and Transport in Terrestrial and Aquatic Ecosystems, a Critical Review of Data, vol. 2. A.A. Balkema, Rotterdam.
Coughtrey, P.J., Thorne, M.C., 1983b. Radionuclide Distribution and Transport in Terrestrial and Aquatic Ecosystems, a Critical Review of Data, vol. 1. A.A. Balkema, Rotterdam.
Coughtrey, P.J., Thorne, M.C., 1983c. Radionuclide Distribution and Transport in Terrestrial and Aquatic Ecosystems – a Critical Review of Data, vol. 2. A.A. Balkema, Rotterdam.

Coughtrey, P.J., Jackson, D., Thorne, M.C., 1983. Radionuclide Distribution and Transport in Terrestrial and Aquatic Ecosystems, vol. 3. A.A. Balkema, Rotterdam.

Coughtrey, P.J., Jackson, D., Jones, C.H., et al., 1984. Radionuclide Distribution and Transport in Terrestrial and Aquatic Ecosystems: A Critical Review of Data, vol. 5. A.A. Balkema, Rotterdam.

Crossley, D.A.J., 1961. Movement and Accumulation of Radiostrontium and Radiocesium in Insects. In: Radioecology. Reinhold Publishing Co-operation and The American Institute of Biological Sciences, Fort Collins, CO.

Crossley, D.A.J., 1973. Comparative movement of Ru-106, Co-60 and Cs-137 in arthropod food chains. In: Nelson, D.J., Evans, F.C. (Eds.), Symposium on Radioecology. Proceedings of the Second National Symposium. USAEC, Washington. United States Atomic Energy Commission (USAEC Report, Conf – 670503), pp. 687–695.

Culioli, J-L., Fouquoire, A., Calendini, S., et al., 2009. Trophic transfer of arsenic and antimony in a freshwater ecosystem: a field study. Aquat. Toxicol. 94, 286–293.

Deitermann, W.I., Hauschild, J., Robenspalavinskas, E., et al., 1989. Soil-to-vegetation transfer of natural I-127, and of I-129 from global fallout, as revealed by field-measurements on permanent pastures. J. Environ. Radioact. 10, 79–88.

Dragovic, S., Jankovic Mandic, L.J., 2010. Transfer of radionuclides to ants, mosses and lichens in semi-natural ecosystems. Radiat. Environ. Biophys. 49, 625–634.

Dragovic, S., Howard, B.J., Caborn, J.A., et al., 2010. Transfer of natural and anthropogenic radionuclides to ants, bryophytes and lichen in a semi-natural ecosystem. Environ. Monitor. Assess. 166, 667–686.

Dubynin, I.D., 1987. Migration of 129I in a freshwater ecosystem. Ekologiya 5, 91–92.

Dushauskene-Duzh, N.-R.F., 1969. A Comparative Study into Accumulation of Strontium-90 and Lead-210 in Fresh Water Hydrobionts of the Lithuanian Republic. Ph.D. Thesis. INUM, Sebastopol (in Russian).

Edgington, D.N., Wahlgren, M.A., Marshall, J.S., 1976. The behaviour of plutonium in aquatic ecosystems: a summary of studies on the Great Lakes. In: Miller, M.W., Stannard, J.N. (Eds.), Environmental Toxicity of Aquatic Radionuclides: Models and Mechanisms. Ann Arbor Science Publishers, Ann Arbor, Michigan, pp. 45–79.

Efroymson, R.A., Sample, B.E., Suter, G.W., 2001. Uptake of inorganic chemicals from soil by plant leaves: regressions of field data. Environ. Toxicol. Chem. 20, 2561–2571.

Emery, R.M., Klopfer, D.C., Garland, T.R., et al., 1976. The Ecological Behaviour of Plutonium and Americium in a Freshwater Ecosystem. PNL Annual Report for 1975, Pt. 2, Ecological Sciences. BNWL-2000. Battelle Pacific Northwest Laboratories, Washington.

Engdahl, A., Ternsell, A., Hannu, S., 2006. Oskarshamn site investigation – Chemical characterisation of deposits and biota. SKB report P-06-320. Swedish Nuclear Fuel and Waste Management Co, Stockholm.

Eyman, L.D., Trabalka, J.R., 1980. Patterns of transuranic uptake by aquatic organisms: consequences and implications. In: Hanson, W.C. (Ed.), Transuranic Elements in the Environment. Technical Information Center, US Department of Energy, Oak Ridge TN, USA, pp. 612–624.

Ferenbaugh, J.K., Fresquez, P.R., Ebinger, M.H., et al., 2002. Radionuclides in soil and water near a low level disposal site and potential ecological and human health impacts. Environ. Monitor. Assess. 74, 243–254.

Folsom, T.R., Wong, K.M., Hodge, V.F., 1973. Some extreme accumulations of natural polonium radioactivity observed in certain oceanic organisms. In: The Natural Radiation Environment II, August 7-11, 1972, Houston, Texas. USERDA, Report. CONF-720805. pp. 863–882.

Garten, C.T., Trabalka, J.R., Bogle, M.A., 1981. Comparative food chain behaviour and distribution of actinide elements in and around a contaminated fresh-water pond. In: International Symposium on Migration in the Terrestrial Environment of Long-Lived Radionuclides from the Nuclear Fuel Cycle. July 1981, Knoxville, Tn, pp. 12–24.

Gashchak, S., Beresford, N.A., 2009. Data forwarded through N. Beresford (personal communication).

Gaso, I., Segovia, N., Morton, O., 2002. In situ biological monitoring of radioactivity and metal pollution in terrestrial snails Helix aspersa from a semiarid ecosystem. Radioprotection 37, 865–871.

Gaso, M.I., Segovia, N., Morton, O., 2005. Environmental impact assessment of uranium ore mining and radioactive waste around a storage centre from Mexico. Radioprotection 40, S739–S745.

Ghuman, G.S., Motes, B.G., Fernandez, S.J., et al., 1993. Distribution of antimony-125, cesium-137 and iodine-129 in the soil-plant system around a nuclear fuel reprocessing plant. J. Environ. Radioact. 21, 161–176.

Giles, M.S., Twinning, J.R., Williams, A.R., et al., 1990. Final report of the Technical Assessment Group for the Maralinga rehabilitation project. Study no. 2. Radioecology. In: Rehabilitation of Former Nuclear Test Sites in Australia. AGPS, Canberra, p. 1.

Gilhen, M., 2001. Current Radionuclide Activity Concentrations in the Chernobyl Exclusion Zone and an Assessment of the Ecological Impact. M.Sc. Thesis. University of Liverpool, Liverpool.

Golder Associates Ltd., 2005. Distribution of Metals in the Aquatic Environment at the Key Lake and McArthur River Operations. Cameco Corporation, Saskatoon, Saskatchewan.

Golubev, A.P., Sikorski, V.G., Kalinin, V.N., et al., 2007. The radioactive contamination dynamics of water body ecosystems of different types in the Chernobyl atomic station alienation zone. Radiobiologiya 47, 326–329.

Gomez, L.S., Marietta, M.G., Jackson, D.W., 1991. Compilation of selected marine radioecological data for the Formerly Utilized Sites Remedial Action Program: Summaries of available radioecological concentration factors and biological half-lives. Sandia National Laboratories Report SAND89-1585 RS-8232-2. Sandia National Laboratories, Albuquerque, USA.

Graham, R.V., Blaylock, B.G., Hoffman, F.O., et al., 1992. Comparison of selenomethionine and selenite cycling in freshwater experimental ponds. Water Air Soil Pollut. 62, 25–42.

Green, N., Hammond, D.J., Davidson, M.F., et al., 2002. The Radiological Impact of Naturally Occurring Radionuclides in Foods from the Wild. NRPB-W30. National Radiological Protection Board, Didcot.

Gudkov, D.I., Derevets, V.V., Zub, L.N., et al., 2005. The distribution of the radionuclides in the main components of lake ecosystems within the Chernobyl NPP exclusion zone. Radiacion. Biol. Radioekol. 45, 271–280.

Guthrie, R.K., Davis, E.M., Cherry, D.S., et al., 1979. Biomagnification of heavy metals by organisms in the marine microcosm. Bull. Environ. Contam.Toxicol. 31, 53–61.

Hameed, P.S., Asokan, R., Iyengar, M.A.R., et al., 1993. The freshwater mussel Parreysia favidens (Benson) as a biological indicator of polonium-210 in a riverine system. Chem. Ecol. 8, 11–18.

Hameed, P.S., Shaheed, K., Somasundaram, S.S.N., 1997. A study on distribution of natural radionuclide polonium-210 in a pond ecosystem. J. Biosci. 22, 627–634.

Hanson, W.C., 1980. Transuranic elements in arctic tundra ecosystems. In: Hanson, W.C. (Ed.), Transuranic Elements in the Environment. DOE/TIC-22800. U.S Department of Energy, Washington, DC, pp. 441–458.

Haschek, W.M., Lisk, D.J., Stehn, R.A., 1979. Accumulation of lead in rodents from two old orchard sites in New York. In: Nielsen, S.W., Migaki, G., Scarpelli, D.G. (Eds.), Animals as Monitors of Environmental Pollutants. National Academy of Sciences, Washington, DC, pp. 192–199.

Hendriks, A.J., Ma, W.C., Brouns, J.J., et al., 1995. Modelling and monitoring organochlorine and heavy metal accumulation in soils, earthworms, and shrews in Rhine-delta floodplains. Arch. Environ. Contam. Toxicol. 29, 115–127.

Hewett, C.J., Jefferies, D.F., 1978. The accumulation of radioactive caesium from food by the plaice (Pleuronectes platessa) and the brown trout (Salmo trutta). J. Fish Biol. 13, 143–153.

Heyraud, M., Cherry, R.D., 1979. Polonium-210 and lead-210 in marine food chains. Mar. Biol. 52, 227–236.

Higley, K.A., 2010. Estimating transfer parameters in the absence of data. Radiat. Environ. Biophys. 49, 645–656.

Higley, K.A., Domotor, S.L., Antonio, E.J., 2003. A kinetic–allometric approach to predicting tissue radionuclide concentrations for biota. J. Environ. Radioact. 66, 61–74.

Holm, E., Ballestra, S., Lopez, J.J., et al., 1994. Radionuclides in macro algae at Monaco following the Chernobyl accident. J. Radioanalyt. Nucl. Chem. 177, 51–72.

Holm, E., Persson, R.B.R., 1980. Behaviour of natural (Th, U) and artificial (Pu, Am) actinides in coastal waters. In: Marine Radioecology, Proceedings of the 3rd NEA Seminar, Tokyo, 1–5 October 1979, OECD, Paris, pp. 237–244.

Hope, B., Loy, C., Miller, P., 1996. Uptake and trophic transfer of barium in a terrestrial ecosystem. Bull. Environ. Contam. Toxicol. 56, 683–689.

Hosseini, A., Thørring, H., Brown, J.E., Saxén, R., Ilus, E., 2008. Transfer of radionuclides in aquatic ecosystems – default concentration ratios for aquatic biota in the Erica tool. J. Environ. Radioact. 99, 1408–1429.

Hou, X., Yan, X., 1998. Study on the concentration and seasonal variation of inorganic elements in 35 species of marine algae. Sci. Total Environ. 222, 141–156.

Hunter, B.A., Johnson, M.S., 1984. Food chain relationship of copper and cadmium in herbivorous and insectivorous small mammals. In: Osborn, D. (Ed.), Metals in Animals. CEH Monkswood, Huntingdon, pp. 5–10.

Hussein, M.A., Obuid-Allah, A.H., Mohammad, A.H., et al., 2006. Seasonal variation in heavy metal accumulation in subtropical population of the terrestrial isopod, Porcellio laevis. Ecotoxicol. Environ. Saf. 63, 168–174.

IAEA, 2004. Sediment Distribution Coefficients and Concentration Factors for Biota in the Marine Environment. IAEA Technical Reports Series No. 422. International Atomic Energy Agency, Vienna.

IAEA, 2010. Handbook of Parameter Values for the Prediction of Radionuclide Transfer in Terrestrial and Freshwater Environments. IAEA Technical Report Series No. 472. International Atomic Energy Agency, Vienna.

IAEA, in preparation. Handbook of Parameter Values for the Prediction of Radionuclide Transfer to Wildlife. IAEA Technical Report Series. International Atomic Energy Agency, Vienna.

Jinks, S.M., Eisenbud, M., 1972. Concentration factors in aquatic environment. Radiat. Data Rep. 13, 243.

Johnson, M.S., Roberts, R.D., 1978. Distribution of lead, zinc and cadmium in small mammals from polluted environments. OIKOS 30, 153–159.

Jones, K.C., Peterson, P.J., Davies, B.E., et al., 1985. Determination of silver in plants by flameless atomic-absorption spectrometry and neutron-activation analysis. Int. J. Environ. Analyt. Chem. 21, 23–32.

Kahn, B., 1980. The bioaccumulation factor for phosphorus-32 in edible fish tissue. U.S. Nuclear Regulatory Commission; NUREG/CR-1336. NTIS, Springfield, VA, University of Michigan Library, p. 116.

Kevern, N.R., Spigarelli, S.A., 1971. Effects of selected limnological factors on the accumulation of cesium-137 fallout by largemouth bass (Micropterus salmoides). Proceedings of the Third National Symposium on Radioecology, 10–12 May 1971, Oak Ridge, TN, pp. 354–360.

Krumholz, L.A., 1956. Observations on the food population of a lake contaminated by radioactive wastes. Bull. Am. Museum Nat. Hist. 110, 277–368.

Kryshev, A.I., Ryabov, I.N., 2005. Model for calculation of fish contamination by Cs-137 and its application for lake Kozhanovskoe (Bryansk region). Radiacion. Biol. Radioekol. 45, 338–345.

Kulikov, N.V., Chebotina, M.Y.A., 1988. Radioecology of fresh water biosystems. In: Radioecology of Fresh-Water Biosystems. Nauka, Sverdlovsk, pp. 1–129 (in Russian).

Kulikov, N.V., Kulikova, V.G., 1977. On Sr-90 and Cs-137 accumulation by some representatives of freshwater fish in natural conditions. Ekologiya 5, 45–49.

Kulikov, N.V., Molchanova, I.V., 1975. Continental radioecology. In: Soil and Freshwater Ecosystems, Nauka, Moscow (in Russian).

Kumblad, L., Bradshaw, C., 2008. Element composition of biota, water and sediment in the Forsmark area, Baltic Sea. Concentrations, bioconcentration factors and partitioning coefficients (Kd) of 48 elements, Technical report, SKB TR-08-09, Swedish Nuclear Fuel and Waste Management Co, Stockholm. 109 pp. Available from: <http://www.skb.se> (Accessed January 2011).

Kurabayashi, M., Fukuda, S., Kurokawa, Y., 1980. Concentration Factors of Marine Organisms Used for the Environmental Dose Assessment. In: Marine Radioecology. NEA/OECD, Paris, pp. 335–345.

Lambrechts, A., Foulquier, L., Garnier-Laplace, J., 1992. Natural radioactivity in the aquatic components of the main French rivers. Radiat. Prot. Dosim. 45, 253–256.

Lentsch, J.W., Kneip, T.J., Wren Mcdonald, E., et al., 1973. Stable Mn and Mn-54 distributions in the physical and biological components of the Hudson River Estuary. In: Nelson, D.J. (Ed.), Radionuclides in Ecosystems. CONF-710501, NITS, Springfield, Virginia, pp. 752–768.

Lindner, G., Becker, M., Eckmann, R., et al., 1990. Biological transfer and sedimentation of Chernobyl radionuclides in Lake Constance. In: Tilzer, M.M., Serruya, C. (Eds.), Large Lakes: Ecological Structure and Function. Brock/Springer Series in Contemporary BioScience. Springer Verlag, Berlin, pp. 265–287.

Little, C.A., 1980. Plutonium in a grassland ecosystem. In: Hanson, W.C. (Ed.), Transuranic Elements in the Environment. DOE/TIC-22800. U.S Department of Energy, Washington, DC, pp. 420–440.

Locatelli, C., Torsi, G., 2000. Determination of Se, As, Cu, Pb, Cd, Zn an Mn by anodic and cathodic stripping voltammetry in marine environmental matrices in the presence of reciprocal interference. Proposal of a new analytic procedure. Microchem. J. 65, 293–303.

Lowson, R.T., Williams, A.R., 1985. A baseline radioecological survey, Manyingee Uranium Prospect, Western Australia. Australian Atomic Energy Commission, Research Establishment, Lucas Heights Research Laboratories, Lucas Heights, New South Wales, Australia.

Malik, N., Biswas, A.K., Qureshi, T.A., et al., 2010. Bioaccumulation of heavy metals in fish tissues of a freshwater lake of Bhopal. Environ. Monitor. Assess. 160, 267–276.

Markham, O.D., Puphal, K.W., Filer, T.D., 1978. Radionuclides in soil and water near a low level disposal site and potential ecological and human health impacts. Environ. Monitor. Assess. 7, 422–428.

Marshall, J.S., Wailerand, B.J., Yaguchi, E.M., 1975. Plutonium in the Laurentian Great Lakes: food-chain relationship. Verh. Internat. Verein. Limnol. 19, 323–329.

Martin, P., Hancock, G.J., Johnston, A., et al., 1995. Bioaccumulation of Radionuclides in Traditional Aboriginal Foods from the Magela and Cooper Creek Systems. Research Report 11. Supervising Scientist for the Alligator Rivers Region, Australian Government Publishing Services, Canberra, pp. 1–53.

Martin, P., Hancock, G.J., Johnston, A., et al., 1998. Natural-series radionuclides in traditional North Australian Aboriginal foods. J. Environ. Radioact. 40, 37–58.

Mcdonald, P., Cook, G.T., Baxter, M.S., 1992. Natural and anthropogenic radioactivity in coastal regions of the UK. Radiat. Prot. Dosim. 45, 707–710.

Meinhold, A.F., Hamilton, L.D., 1992. Radium concentration factors and their use in health and environmental risk assessment. In: Ray, J.P., Engelhardt, F.R. (Eds.), Produced Water Technological/ Environmental Issues and Solutions. Plenum Press, New York, pp. 293–302.

Mietelski, J.W., 2001. Plutonium in the environment of Poland (a review). In: Kudo, A. (Ed.), Plutonium in the Environment. Elsevier, Amsterdam, pp. 401–412.

Mietelski, J.W., Swalko, P., Tomankiewicz, E., et al., 2004. Cs-137, K-40, Sr-90, Pu-238, Pu-239 + 240, Am-241 and Cm243 + 244 in forest litter and their transfer to some species of insects and plants in boreal forests: three case studies. J. Radioanalyt. Nucl. Chem. 262, 645–660.

Mohamed, F.A.S., 2008. Bioaccumulation of selected metals and histopathological alterations in tissues of Oreochromis niloticus and Lates niloticus from Lake Nasser, Egypt. Global Veterin. 2, 205–218.

Nagy, K.A., 2001. Food requirements of wild animals: predictive equations for free-living mammals, reptiles and birds. Nutr. Abs. Rev. Ser. B 71, 21–31.

Newman, M.C., Brisbin Jr., I.L., 1990. Variation of Cs-137 levels between sexes, body sizes, and collection localities of mosquitofish, Gambusia holbrooki (Girard, 1859), inhabiting a reactor cooling reservoir. J. Environ. Radioact. 12, 131–141.

Nilsson, M., Dahlgaard, H., Edgren, M., et al., 1981. Radionuclides in Fucus from inter-Scandinavian waters. In: Impacts Radionuclides in Fucus from Inter-Scandinavian Waters. Impacts of Radionuclide Releases into the Marine Environment, Proceedings Symposium, Vienna, 6–10 October 1980. IAEA-SM-248/107, Vienna, p. 5.

NIRS, 2009. Studies on the Environmental Transfer Parameters of Radionuclides in the Japanese Biosphere. NIRS Annual, Ministry of Economy, Trade and Industry (METI), Chiba City.

Ophel, I.L., Fraser, J.M., Judd, J.M., 1972. Concentration factors and bottom sediments of a freshwater lake. In: Radioecology Application to Protection of Man and his Environment, Commission of the European Communities. Luxembourg, pp. 509–530.

Opydo, J., Ufnalski, K., Opydo, W., 2005. Heavy metals in Polish forest stands of Quercus robur and Q. petraea. Water Air Soil Pollut. 161, 175–192.
Outola, I., Saxen, R., Heinävaara, S., 2009. Transfer of Sr-90 into fish in Finnish lakes. J. Environ. Radioact. 100, 657–664.
Ozturk, M., Ozozen, G., Minareci, O., et al., 2009. Determination of heavy metals in fish, water and sediments of Avsar dam lake in Turkey. Iran. J. Environ. Health Sci. Eng. 6, 73–80.
Panchenko, S.V., Panfilova, A.A., 2000. Regarding the role of forest ecosystems in exposure of the population. In: Panchenko S.V. (Ed.), Problems of Forest Radioecology. MOGUL, Moscow, pp. 228–293 (in Russian).
Pentreath, R.J., 1981. The presence of 237Np in the Irish Sea. Mar. Ecol. Progr. Ser. 6, 243–247.
Peterson, L.R., Trivett, V., Baker, A.J.M., et al., 2003. Spread of metals through an invertebrate food chain as influenced by a plant that hyperaccumulates nickel. Chemoecology 13, 103–108.
Pokarzhevskii, A., Zhulidov, A., 1995. Halogens in soil animal bodies: a background level. In: Van den Brink, W.J., Bosman, R., Arendt, F. (Eds.), Contaminated Soil. Academic Publishers, Dordrecht, pp. 403–404.
Pokarzhevskii, A.D., Krivolutzkii, D.A., 1997. Background concentrations of Ra-226 in terrestrial animals. Biogeochemistry 39, 1–13.
Polikarpov, G.G., 1966. Radioecology of Aquatic Organisms: the Accumulation and Biological Effects of Radioactive Substances. Reinhold, New York.
Porntepkasemsan, B., Nevissi, A.E., 1990. Mechanism of radium-226 transfer from sediments and water to marine fishes. Geochem. J. 24, 223–228.
Poston, T.M., Klopfer, D.C., 1986. A Literature Review of the Concentration Ratios of Selected Radionuclides in Freshwater and Marine Fish. PNL-5484. Pacific Northwest Laboratory, Richland, WA.
Preston, D.F., Dutton, J.W.R., 1967. The concentrations of caesium-137 and strontium-90 in the flesh of brown trout taken from rivers and lakes in the British Isles between 1961 and 1966: The variables determining the concentrations and their use in radiological assessments. Water Res. 1, 475–496.
Rashed, M.N., 2001. Monitoring of environmental heavy metals in fish from Nasser Lake. Environ. Int. 27, 27–33.
Read, H.J., Martin, M.H., 1993. The effect of heavy metals on populations of small mammals from woodlands in Avon (England); with particular emphasis on metal concentrations in Sorex araneus L. and Sorex minutus L. Chemosphere 27, 2197–2211.
Read, J., Pickering, R., 1999. Ecological and toxicological effects of exposure to an acid, radioactive tailings storage. Environ. Monitor. Assess. 54, 69–85.
RIFE, 1996. RIFE 1, Radioactivity in Food and the Environment 1995. Ministry of Agriculture, Fisheries & Food. Available from: <http://www.sepa.org.uk/radioactive_substances/publications/rife_reports.aspx> (Accessed August 2011).
RIFE, 1997. RIFE 2, Radioactivity in Food and the Environment 1996. Ministry of Agriculture, Fisheries & Food, Scottish Environment Protection Agency. Available from: <http://www.sepa.org.uk/radioactive_substances/publications/rife_reports.aspx> (Accessed August 2011).
RIFE, 1998. RIFE 3, Radioactivity in Food and the Environment 1997. Ministry of Agriculture, Fisheries & Food, Scottish Environment Protection Agency. Available from: <http://www.sepa.org.uk/radioactive_substances/publications/rife_reports.aspx> (Accessed August 2011).
RIFE, 1999. RIFE 4, Radioactivity in Food and the Environment 1998. Ministry of Agriculture, Fisheries & Food, Scottish Environment Protection Agency. Available from: <http://www.sepa.org.uk/radioactive_substances/publications/rife_reports.aspx> (Accessed August 2011).
RIFE, 2000. RIFE 5, Radioactivity in Food and the Environment, 1999. Food Standards Agency, Scottish Environment Protection Agency. Available from: <http://www.sepa.org.uk/radioactive_substances/publications/rife_reports.aspx> (Accessed August 2011).
RIFE, 2001. RIFE 6, Radioactivity in Food and the Environment 2000. Food Standards Agency, Scottish Environment Protection Agency. Available from: <http://www.sepa.org.uk/radioactive_substances/publications/rife_reports.aspx> (Accessed August 2011).

RIFE, 2002. RIFE 7, Radioactivity in Food and the Environment 2001. Food Standards Agency, Scottish Environment Protection Agency. Available from: <http://www.sepa.org.uk/radioactive_substances/publications/rife_reports.aspx> (Accessed August 2011).

RIFE, 2003. RIFE 8, Radioactivity in Food and the Environment 2002. Environment Agency, Environment and Heritage Service, Food Standards Agency, Scottish Environment Protection Agency. Available from: <http://www.sepa.org.uk/radioactive_substances/publications/rife_reports.aspx> (Accessed August 2011).

RIFE, 2004. RIFE 9, Radioactivity in Food and the Environment 2003. Environment Agency, Environment And Heritage Service, Food Standards Agency, Scottish Environment Protection Agency. Available from: <http://www.sepa.org.uk/radioactive_substances/publications/rife_reports.aspx> (Accessed August 2011).

Roberts, R.D., Johnson, M.S., Hutton, M., 1978. Lead contamination of small mammals from abandoned metalliferous mines. Environ. Pollut. 15, 61–69.

Rowan, D.J., Rasmussen, J.B., 1994. Bioaccumulation of radiocesium by fish – the influence of physicochemical factors and trophic structure. Can. J. Fish. Aquat. Sci. 51, 2388–2410.

Ryabokon, N.I., Smolich, I.I., Kudryashov, V.P., et al., 2005. Long-term development of the radionuclide exposure of murine rodent populations in Belarus after the Chernobyl accident. Radiat. Environ. Biophys. 44, 169–181.

Ryan, B., Bollhöfer, A., Medley, P., 2009. Bioaccumulation in Terrestrial Plants on Rehabilitated Landforms. ERISS Research Summary 2007–2008. NT152-159. In: Jones, D.R., Webb, A. (Eds.), Supervising Science Report 200. Supervising Scientist, Darwin.

Sample, B.E., Suter, G.W., 2002. Screening evaluation of the ecological risks to terrestrial wildlife associated with a coal ash disposal site. Hum. Ecol. Risk Assess. 8, 637–656.

Saxén, R., Outola, I., 2009. Polonium-210 in freshwater and brackish environment. In: Gjelsvik R., Brown, J.E. (Eds.), A Deliverable Report for the NKS-B Activity October 2008, GAPRAD – Filling Knowledge Gaps in Radiation Protection Methodologies for Non-human Biota. Nordic Nuclear Safety Research (NKS), DK - 4000 Roskilde, Denmark, pp. 13–21.

Scheuhammer, A.M., Bond, D.E., Burgess, N.M., et al., 2003. Lead and stable lead isotope ratios in soil, earthworms, and bones of American woodcock (Scolopax minor) from eastern Canada. Environ. Toxicol. Chem. 22, 2585–2591.

Shaheed, K., Somasundaram, S.S.N., Hameed, P.S., et al., 1997. A study of polonium-210 distribution aspects in the riverine ecosystem of Kaveri, Tiruchirappalli, India. Environ. Pollut. 95, 371–377.

Sharma, R.P., Shupe, J.L., 1977. Trace metals in ecosystems: relationships of residues of copper, molybdenum, selenium, and zinc in animal tissues to those in vegetation and soil in the surrounding environment. In: Drucker, H., Wildung, R.E. (Eds.), Biological Implications of Metals in the Environment. CONF 750929, National Technical Information Service, Springfield, Virginia, pp. 595–608.

Sheppard, S.C., Evenden, W.G., 1990. Characteristics of plant concentration ratios assessed in a 64-site field survey of 23 elements. J. Environ. Radioact. 11, 15–36.

Shorti, Z.F., Palumbo, R.F., Oldon, P.B., et al., 1969. Uptake of I-131 by biota of Fern Lake, Washington in a laboratory and field experiment. Ecology 50, 979–989.

Sivintsev, Y.U.V., Vakulovsky, S.M., Vasiliev, A.P., et al., 2005. Anthropogenic radionuclides in seas bounding Russia. Radioecological impact of radioactive waste disposal to arctic and far east seas. In: White Book – 2000. IzdaT, Moscow, p. 248 (in Russian).

Skubala, P., Kafel, A., 2004. Oribatid mite communities and metal bioaccumulation in oribatid species (Acari, Oribatida) along the heavy metal gradient in forest ecosystems. Environ. Pollut. 132, 51–60.

Smagin, A.I., 2006. The study of the multifactor anthropogenic influence on the ecosystems of the industrial reservoirs of 'Mayak' PA. Radiacion. Biol. Radioekol. 46, 94–110.

Stanica, F., 1999. Accumulation of different metals in apple tree organs from an unfertilised orchard. In: Gerzabek, M.H. (Ed.), IUR Soil to Plant Transfer Working Group. International Union of Radioecology, Cadarache, France, pp. 96–100.

Suzuki, H., Koyanagi, T., Saiki, M., 1973. Studies on rare elements in seawater and uptake by marine organisms. In: Impacts of Nuclear Releases into the Aquatic Environment, Proceedings of a

Symposium, Otaniemi, 30 June–4 July 1975. IAEASM-198/19, International Atomic Energy Agency, Vienna, pp. 77–91.
Tagami, K., Uchida, S., 2010. Can elemental composition data of crop leaves be used to estimate radionuclide transfer to tree leaves? Radiat. Environ. Biophys. 49, 583–590.
Tagami, K., Uchida, S., 2005. Soil-to-plant transfer factors of technetium-99 for various plants collected in the Chernobyl area. J. Nucl. Radiochem. Sci. 6, 261–264.
Takata, H., Aono, T., Tagami, K., et al., 2010. Concentration ratios of stable elements for selected biota in Japanese estuarine areas. Radiat. Environ. Biophys. 49, 591–601.
Trapeznikov, A., Pozolotina, V., Chebotina, M., et al., 1993a. Radioactive contamination of the Techa River, the Urals. Health Phys. 65, 481–488.
Trapeznikov, A.V., Pozolotina, V.N., Chebotina, M.Y.A., et al., 1993b. Radioactive contamination of the River Techa in the Urals. Ekologiya 6, 72–77.
Trapeznikov, A.V., 2001. Radioecology of Freshwater Ecosystems (Exemplified by the Urals Region). Doctor of Sciences Thesis. Institute of Ecology of Plants and Animals, Ekaterinburg (in Russian).
Trapeznikova, V.N., Trapeznikov, A.V., Kulikov, N.V., 1984. Cs-137 accumulation in food fish of the cooling pond of the Beloyarsk NPP. Ekologiya 4, 36–39.
Trapeznikov, A.V., Molchanova, I.V., Karavaeva, E.N., et al., 2007. Freshwater ecosystems radionuclide migration in freshwater and terrestrial ecosystems. Urals Branch of the Russian Academy of Sciences, Ekaterinburg, pp. 356–357 (in Russian).
UNEP, ILO, WHO, 1983. Environmental Health Criteria 25: Selected Radionuclides – Tritium, Carbon-14, Krypton-85, Strontium-90, Iodine, Caesium-137, Radon and Plutonium. International Programme on Chemical Safety, WHO, Geneva.
Vakulovsky, S.M., 2008. The Radiation Situation Within Russia and Adjacent States in 2007. RosHydromet-SPA Typhoon, Obninsk (in Russian).
Van As, D., Fourie, H.O., Vleggaar, C.M., 1975. Trace element concentrations in marine organisms from the Cape West Coast. S.A.J. Sci. 71, 151–154.
Vanderploeg, H.A., Parzcyk, D.C., Wilcox, W.H., et al., 1975. Bioaccumulation Factors for Radionuclides in Freshwater Biota. ORNL-5002. Oak Ridge National Laboratory, Oak Ridge, TN.
Verhovskaya, I.N., 1972. Radioecological Investigations in Natural Biogeocenoses. Nauka, Moscow (in Russian).
Vermeulen, F., Van Den Brink, N.W., D'Havé, H., et al., 2009. Habitat type-based bioaccumulation and risk assessment of metal and As contamination in earthworms, beetles and woodlice. Environ. Pollut. 157, 3098–3105.
Vetikko, V., Saxen, R., 2010. Application of the ERICA assessment tool to freshwater biota in Finland. J. Environ. Radioact. 101, 82–87.
Vintsukevich, N.V., Tomilin, Y.U.A., 1987. Radionuclide distribution in aquatic system (NPP cooling pond-river-sea estuary). Ekologiya 6, 72.
Whicker, F.W., Little, C.A., Winsor, T.F., 1974. Plutonium behaviour in the terrestrial environs of the Rocky Flats installation. Environmental Surveillance Around Nuclear Installations, 5–9 November 1973, Warsaw. IAEA-SM-180/4589-103, International Atomic Energy Agency, Vienna.
Williams, A.R., 1978. The distribution of some naturally occurring elements in the environment of the Yeelirrie Uranium Deposit, Western Australia. In: Brownscombe, A.J., Davy, D.R., Giles, M.S., Williams, A.R. (Eds.), Three Baseline Studies in the Environment of the Uranium Deposit at Yeelirrie. Australian Atomic Energy Commission, Sydney.
Williams, A.R., 1981. Background Radiological Data for the Proposed Beverly Uranium Development, South Australia. Australian Atomic Energy Commission, Research Establishment, Sydney.
Wood, M.D., 2010. Assessing the Impact of Ionising Radiation in Temperate Coastal Sand Dunes: Measurement and Modelling. Ph.D. Thesis. University of Liverpool, Liverpool.
Wood, M.D., Leah, R.T., Jones, S.R., et al., 2009. Radionuclide transfer to invertebrates and small mammals in a coastal sand dune ecosystem. Sci. Total Environ. 407, 4062–4074.
Wood, M.D., Beresford, N.A., Semenov, D.V., et al., 2010. Radionuclide transfer to reptiles. Radiat. Environ. Biophys. 49, 509–530.
Yankovich, T.L., 2010. Compilation of Concentration Ratios for Aquatic Non-human Biota Collected by the Canadian Power Reactors Sector. CANDU Owners Group Inc., p. 7.

Yankovich, T.L., Sharp, K.J., Benz, M.L., Carr, J., Killey, R.W.D., 2008. Carbon-14 specific activity model validation for biota in wetland environments. In: Proceedings of the ANS Topical Meeting on Decommissioning, Decontamination, and Reutilization, 16–19 September 2007, Chattanooga, TN.

Yoshida, S., Muramatsu, Y., Peijnenburg, W.J.G.M., 2005. Multi-element analyses of earthworms for radioecology and ecotoxicology. Radioprotection 40, S491–S495.

Yu, C., 2007. Modeling radionuclide transport in the environment and assessing risks to humans, flora, and fauna: the RESRAD family of codes. In: Applied Modeling and Computations in Nuclear Science. American Chemical Society, Washington, DC, USA, pp. 58–70 (Chapter 5).

Zesenko, A.Y.A., Kulebyakina, L.G., 1982. Sr-90 content in the Danube mouth and adjacent north-western part of the Black Sea. Ekologiya 5, 39.

ANNEX C. SELECTED DATA FOR REFERENCE FLATFISH

Table C.1. Internal body distributions of selected elements in Reference Flatfish: data for the European Plaice – *Pleuronectes platessa L.*

Element	Tissue	Concentration ratio fresh weight (Bq/kg per Bq/l)	Notes	Reference
Zn	Whole blood	600	Pooled sample of five plaice (each weighing 30 g), 15 μg Zn/l unfiltered seawater	Pentreath (1973a)
	Blood cells	833		
	Blood plasma	733		
	Heart	1173		
	Spleen	1587		
	Liver	1727		
	Kidney	1973		
	Gonad	8853		
	Gut	1253		
	Gill	2180		
	Skin	2480		
	Muscle	373		
	Bone	2180		

Element	Tissue	Concentration ratio fresh weight (Bq/kg per Bq/l)	Notes	Reference
Mn	Whole blood	35	Pooled sample of five plaice (each weighing 30 g), 2 μg Mn/l unfiltered seawater	Pentreath (1973a)
	Blood cells	2750		
	Heart	275		
	Liver	695		
	Kidney	360		
	Gonad	10		
	Gut	365		
	Gill	385		
	Skin	1545		
	Muscle	110		
	Bone	9090		

Element	Tissue	Concentration ratio fresh weight (Bq/kg per Bq/l)	Notes	Reference
Co	Whole blood	167	Pooled sample of five plaice (each weighing 30 g), 0.25 μg Co/l unfiltered seawater	Pentreath (1973b)
	Spleen	1125		
	Liver	2042		
	Gut	250		
	Gill	375		
	Skin	4042		
	Muscle	83		
	Bone	5250		

Element	Tissue	Concentration ratio fresh weight (Bq/kg per Bq/l)	Notes	Reference
Cd	Blood cells	<50	Pooled sample of five plaice (each weighing 45 g), 0.1 μg Cd/l unfiltered seawater	Pentreath (1977a)
	Heart	50		
	Spleen	300		
	Liver	2300		
	Kidney	460		
	Gut	480		
	Gill filaments	200		
	Skin	680		
	Muscle	100		
	Bone	<1000		

Element	Tissue	Concentration ratio fresh weight (Bq/kg per Bq/l)	Notes	Reference
Ag	Muscle	250	Pooled sample of five plaice, 0.04 μg Ag/l unfiltered seawater	Pentreath (1977b)
	Liver	1200		
	Bone	<500		
	Gut/viscera	600		
	Other (av. blood plasma, spleen, gill, and skin)	712.5		

Table C.2. Accumulation of some radionuclides by eggs of the European Plaice: *Pleuronectes platessa L.*

Radionuclide	Concentration ratio at hatching	Reference
^{90}Sr	0.1	Woodhead (1970)
^{137}Cs	1	Woodhead (1970)
^{54}Mn	2	Pentreath (1976)
^{90}Y	10	Woodhead (1970)
^{106}Ru	10	Woodhead (1970)
^{144}Ce	10	Woodhead (1970)
^{95}Zr	10	Woodhead (1970)
^{95}Nb	10	Woodhead (1970)
^{65}Zn	30	Pentreath (1976)
^{51}Cr	30	Pentreath (1977c)
^{241}Am	25	Pentreath (1977c)
^{239}Pu	35	Pentreath (1977c)
^{110m}Ag	1600	Pentreath (1977b)

Table C.3. Some biological half-times for the European Plaice: *Pleuronectes platessa L.*

Radionuclide	Source of uptake	$Tb_{0.5}$ (days)	Reference
^{131}I	Food	19	Pentreath (1977c)
^{137}Cs	Water	65	Jefferies and Hewett (1971)
^{54}Mn	Water	153	Pentreath (1973a)
^{54}Mn	Food	40	Pentreath (1976)
^{65}Zn	Water	295	Pentreath (1973a)
^{65}Zn	Food	103	Pentreath (1976)
^{59}Fe	Water	105	Pentreath (1973b)
^{58}Co	Water	65	Pentreath (1973b)
^{110m}Ag	Water	31	Pentreath (1977b)
^{110m}Ag	Food	12	Pentreath (1977b)

References

Jefferies, D.F., Hewett, C.J., 1971. The accumulation and excretion of radioactive caesium by the plaice and the thornback ray. J. Mar. Biol. Ass. UK 51, 411–422.

Pentreath, R.J., 1973a. The accumulation of ^{65}Zn and ^{54}Mn by the plaice, Pleuronectes platessa L. J. Exp. Mar. Biol. Ecol. 12, 1–18.

Pentreath, R.J., 1973b. The accumulation of ^{59}Fe and ^{58}Co by the plaice, Pleuronectes platessa L. and the thornback ray, Raja clavata L. J. Exp. Mar. Biol. Ecol. 12, 315–326.

Pentreath, R.J., 1976. Some further studies on the accumulation and retention of ^{65}Zn and ^{54}Mn by the plaice, Pleuronectes platessa L. J. Exp. Mar. Biol. Ecol. 21, 179–189.

Pentreath, R.J., 1977a. The accumulation of cadmium by the plaice, Pleuronectes platessa L. and the thornback ray, Raja clavata L. J. Exp. Mar. Biol. Ecol. 30, 223–232.

Pentreath, R.J., 1977b. The accumulation of ^{110m}Ag by the plaice, Pleuronectes platessa L. and the thornback ray, Raja clavata L. J. Exp. Mar. Biol. Ecol. 29, 315–325.

Pentreath, R.J., 1977c. Radionuclides in marine fish. Oceanogr. Mar. Biol. Ann. Rev. 15, 365–460.

Woodhead, D.S., 1970. The assessment of the radiation dose to developing fish embryos due to the accumulation of radioactivity by the egg. Radiat. Res. 43, 582–597.

Annals of the ICRP

Published on behalf of the International Commission on Radiological Protection

---

Aims and Scope

The International Commission on Radiological Protection (ICRP) is the primary body in protection against ionising radiation. ICRP is a registered charity and is thus an independent non-governmental organisation created by the 1928 International Congress of Radiology to advance for the public benefit the science of radiological protection. The ICRP provides recommendations and guidance on protection against the risks associated with ionising radiation, from artificial sources widely used in medicine, general industry and nuclear enterprises, and from naturally occurring sources. These reports and recommendations are published approximately four times each year on behalf of the ICRP as the journal *Annals of the ICRP*. Each issue provides in-depth coverage of a specific subject area.

Subscribers to the journal receive each new report as soon as it appears so that they are kept up to date on the latest developments in this important field. While many subscribers prefer to acquire a complete set of ICRP reports and recommendations, single issues of the journal are also available separately for those individuals and organizations needing a single report covering their own field of interest. Please order through your bookseller, subscription agent, or direct from the publisher.

ICRP is composed of a Main Commission, a Scientific Secretariat, and five standing Committees on: radiation effects, doses from radiation exposure, protection in medicine, the application of ICRP recommendations, and protection of the environment. The Main Commission consists of a Chair and twelve other members. Committees typically comprise 10–15 members. Biologists and medical doctors dominate the current membership; physicists are also well represented.

ICRP uses Working Parties to develop ideas and Task Groups to prepare its reports. A Task Group is usually chaired by an ICRP Committee member and usually contains a number of specialists from outside ICRP. Thus, ICRP is an independent international network of specialists in various fields of radiological protection. At any one time, about one hundred eminent scientists and policy makers are actively involved in the work of ICRP. The Task Groups are assigned the responsibility for drafting documents on various subjects, which are reviewed and finally approved by the Main Commission. These documents are then published as the *Annals of the ICRP*.

The membership of the Task Group that prepared this report was:

Full Members
P. Strand (Chairman)
N. Beresford
D. Copplestone
J. Godoy
L. Jianguo (from 2009)
R. Saxén (to 2009)
T. Yankovich

Corresponding Member
J. Brown